A management guide to birds of lowland farmland

by Richard Winspear and Gethin Davies

CW01662213

Citation

For bibliographic purposes this book should be referred to as
Winspear R and Davies G (2005) *A management guide to birds of lowland farmland*.
The RSPB, Sandy.

The techniques described or referred to in this book do not necessarily reflect the policies of the RSPB or individuals involved in its production.

No responsibility can be accepted for any loss, damage or unsatisfactory results arising from the implementation of any of the recommendations within this book.

The use of proprietory and commercial trade names in this handbook does not necessarily imply endorsement of the product by the authors or publishers.

Published by
The Royal Society for the Protection of Birds
The Lodge, Sandy, Bedfordshire SG19 2DL

Designed by
NatureBureau

Figures by
Dan Powell

Printed by
Information Press

ISBN: 1 901930 57 2

Contributors

John Day	The RSPB	3.3
Nigel Symes	The RSPB	3.19.2
Graham White	The RSPB	3.21, 3.22
Phil Wilson	Wessex Environmental Associates	3.11

Reviewers

Nicholas Aebischer	Game Conservancy Trust	1.8, 2.1
Keith Alexander	Ancient Tree Forum	3.2
Guy Anderson	The RSPB	1.1, 2.5
Malcolm Ausden	The RSPB	1.7, 3.4
Mark Bolton	The RSPB	2.2, 3.18
Richard Bradbury	The RSPB	1.4, 2.6, 2.7
Chris Britt	ADAS	3.1
Stephen Browne	Game Conservancy Trust	2.3
Dave Buckingham	The RSPB	1.5, 3.18, 3.19, 3.24
Jill Butler	Woodland Trust	3.2
Morwenna Christian	Defra (RDS)	3.20
John Day	The RSPB	3.2, 3.5
Andy Evans	The RSPB	1.2
Jo Gilbert	The RSPB	3.20
Julian Hughes	The RSPB	1.8
Darren Moorcroft	The RSPB	2.6
Tony Morris	The RSPB	1.3, 1.7, 2.4, 3.17
Will Peach	The RSPB	1.5, 1.8
Allan Perkins	The RSPB	2.9
Peter Robertson	The RSPB	All chapters
Rob Sheldon	The RSPB	3.16
Mark Thomas	The RSPB	1.8
Peter Thompson	Game Conservancy Trust	3.10, 3.12, 3.14, 3.15
Chris Tomson	The RSPB	3.23
Robert Walton	English Nature	3.23
Andy Wight	The RSPB	3.24
Graham White	The RSPB	1.7, 3.4, 3.23

Contents

Appendices

Index

INTRODUCTION

This book provides information on how agricultural changes have affected lowland bird populations, detailed ecology on key species that have declined and practical advice on farmland management that will benefit these and other wildlife.

The first two chapters are largely based on scientific research by a range of conservation and agricultural organisations and published in scientific journals. Key scientific references are given for further reading if the reader wishes to refer to the original research. The species ecology sections end with a summary of management advice specific to the species. Management of specific features is covered in more detail in the last chapter. The third chapter is based on the knowledge and experience of a wide range of farm conservation advisers. It covers all recognised management topics relevant to the conservation of lowland farmland birds, including those funded by agri-environment schemes in the UK. In some cases, references for further reading are given where management details go beyond the scope possible within this guide.

Where management is planned on set-aside or through an agri-environment scheme, then the advice in this book should be read in conjunction with the current set-aside or scheme rules to ensure compliance with their requirements.

The recommended approach to farmland wildlife conservation is to identify the key species and habitat features present on the farm, maintain or restore important habitats that already exist and provide the key requirements of species that are not already met through the creation of new habitats.

Supplements to this management guide will be produced as and when new information becomes available.

1 FACTORS AFFECTING BIRDS ON LOWLAND FARMLAND

1.1 Factors affecting birds on lowland farmland

The main changes in agriculture that have affected birds are loss of mixed farming, increased use of pesticides, improvement of grasslands, changes in cropping, land drainage and loss of field boundary and margin habitats. The effect of changes in agriculture on bird populations has been the loss of quantity or quality of nesting habitat, the decline in abundance or availability of food, the reduction in opportunities through a loss of habitat diversity, and habitat changes that have increased species' vulnerability to predation.

This chapter will review the changes in land management on farmland that have influenced the populations of lowland farmland birds since 1950. The link between agricultural changes and declines in bird populations has been proven for a range of species and, in three cases, conservation measures based on the scientific findings have been used to reverse these declines.

Grassland improvement and mechanisation has led to the earlier and more frequent cutting of agricultural grassland, which had dramatic impacts on the nest survival of corncrakes (Plate 1.1). The surviving corncrake population in Scotland is recovering through implementation of payments for farmers to leave early cover for them to nest, delaying mowing in nesting fields and mowing from the inside of the field outwards to prevent corncrakes being trapped and killed by the mower.

Reasons for the decline of the stone-curlew (Plate 1.2) in England included loss of short, unimproved grasslands and spring-sown arable land on which they nest, and nest losses due to agricultural operations. A partial recovery has begun through protection of nests during agricultural operations, provision of fallow nesting plots in areas where late spring-sown crops have largely disappeared and appropriate grassland management where unimproved grassland still occurs within their breeding range.

The cirl bunting (Plate 1.3) is now restricted to a mixed farming area in south Devon. Research showed the importance of weedy stubble fields for seed food through the winter and unimproved grassland or rough grass field margins for insect food in the summer for this species. Conservation of these features within their breeding range has led directly to the recovery of this population.

Chris Gomersall/rspb-images.com

Chris Gomersall/rspb-images.com

Chris Gomersall/rspb-images.com

Plates 1.1, 1.2 and 1.3 Populations of the corncrake (left), stone-curlew (top right) and cirl bunting (bottom right) have all increased in recent years through positive land management by farmers.

For other, more widespread species where links between agricultural changes and population declines have been proven, it has not yet been possible to implement conservation measures on a large enough scale to cause a UK population recovery.

In the case of the grey partridge, poor chick survival related to low levels of insect food was identified as the main cause of population decline. Experimental manipulation has proven that reductions in the use of herbicides and insecticides boosts insect food available to the chicks and, in turn, improves breeding productivity. Predator control was also shown to benefit this species and availability of nesting habitat and winter seed food has been identified as a limiting factor in some areas.

Skylarks nest more successfully and at higher densities in spring-sown cereals than in winter-sown crops. The factors determining breeding success were identified as crop height and density, which restricts access to nest sites and food. In areas where spring sowing is rare, experimental establishment of winter cereal crops with small, undrilled patches has been shown to boost skylark productivity by almost 50%.

In addition, scientific evidence for links between agricultural changes and population declines have been found for a range of other species (Table 1.1).

These same species have shown declines elsewhere in Europe, and the scale of the declines is related to the degree of agricultural intensification. For example, population declines have been greater in the UK, Germany, Denmark and the Netherlands than in eastern European countries.

Land management changes have affected one or more of the following requirements:
- the quantity or quality of the nesting habitat
- the abundance or availability of seed food*
- the abundance or availability of insect food*
- the abundance or availability of soil invertebrates*

Table 1.1 The known causes of decline and strength of scientific evidence for lowland farmland bird species.

Species	Causes of decline	Strength of evidence
Grey partridge	Increased pesticide use and predator populations	Proven by experiment
Corncrake	Improvement of meadows leading to earlier and more frequent cutting	Proven by recovery
Lapwing	Loss of mixed farming, spring cropping and unimproved grassland	Strong evidence
Stone-curlew	Loss of unimproved grassland and spring cropping, and nest destruction during agricultural operations	Proven by recovery
Turtle dove	Loss of summer seed food and scrub	Some evidence
Skylark	Loss of spring cropping and improvement of grasslands	Strong evidence
Yellow wagtail	Loss of suitable habitat for nesting and foraging and nest destruction due to earlier mowing of grassland	Some evidence
Song thrush	Loss of damp foraging areas	Some evidence
Tree sparrow	Loss of seed and insect food	Some evidence
Linnet	Loss of summer seed food	Strong evidence
Yellowhammer	Loss of winter seed food and rough grassland areas, and increased pesticide use	Strong evidence
Cirl bunting	Loss of winter seed food, rough grassland areas and mixed farming	Proven by recovery
Reed bunting	Loss of winter seed food and damp habitats	Some evidence
Corn bunting	Loss of winter seed food and increased use of pesticides (and possibly nest destruction through earlier harvesting)	Strong evidence

- the diversity of habitats (some species need a number of habitats in close proximity)
- protection from predation.

* Birds require food resources to be both abundant and available. Overall food abundance is obvious ly critical, but if that food is in the wrong place or difficult to forage, then birds can suffer food shortages as severe as if food abundance were low. For example, if seed and insect food is available in abundance in field margins, then this may not be accessible to skylarks, which avoid foraging in field margins to evade predation. Also, if there is an abundance of insect food in rank grass margins, then many species may find it difficult to forage successfully in the dense vegetation, resulting in a slow feeding rate, similar to feeding in more accessible habitat with a lower insect abundance.

1.1.1 Nesting habitat

Birds have varying nesting habitat requirements because the structure of the vegetation affects the accessibility and visibility of both the birds and their predators. They can broadly be categorised as open field-nesting species (such as the lapwing, skylark, yellow wagtail and corn bunting) and field boundary-nesting species (such as the grey partridge, turtle dove, linnet and yellowhammer).

The suitability of in-field nesting habitat is generally determined by the sward structure, the degree of nest loss from agricultural operations or livestock trampling and availability of in-field food resources.

- Earlier mowing of grasslands has been identified as the cause of decline for the corncrake and has been implicated in the decline of the skylark and yellow wagtail.
- Increased livestock densities have resulted in waders such as lapwings losing their nests through trampling. The increased grazing pressure produces short, uniform swards that provides insufficient nesting cover for some species. Conversely, the rapid grass growth through use of fertilisers on silage and hay fields creates a sward that is too tall for species such as the lapwing.
- On arable land, the switch from spring to autumn sowing has reduced the nesting habitat available to lapwings and reduced the quality of the nesting and foraging habitat for skylarks.
- The loss of hedgerows caused by field enlargement, and the loss of the associated field margin habitats has reduced the amount of nesting habitat available to the large proportion of farmland birds that nest in the field boundaries.

1.1.2 Seed food

Declining seed-eating bird species on farmland tend to be those that feed almost exclusively on farmland habitats. Similar species that have not shown a decline tend to be species, such as the chaffinch, that can also source their food from other habitats, such as woodland and gardens.

The availability of seed food for birds on farmland has been affected by:

- the increased use and efficiency of herbicides
- changes in crop types with different weed spectra
- a move to more autumn sowing that reduces the area left as stubble through the winter
- more efficient harvesting machinery that leave fewer seeds in the stubble
- exclusion of birds from grain stores and animal food stores to meet hygiene regulations
- more intensive grassland management through reseeding, fertiliser use, more frequent cutting and higher stocking rates that reduce the diversity of seeds produced in meadows and pastures
- the decline in area of field margin habitats
- the loss of arable farming in specialist livestock areas.

The important seeds in the diet of declining farmland birds include those of cereals, polygonums, chickweed and fat hen, and the seeds and plant matter of clover and brassica crops. Declines in the abundance of seeds of these plants are generally found on both intensively managed arable and grassland systems.

1.1.3 Insect food

Insect abundance on arable farmland is affected directly through the increased use of insecticides, but also indirectly through the increased use of herbicides, which remove the food plants of many important insects eaten by birds. Many fungicides also have insecticidal properties. The loss of undersown crops and field margin habitats will have also affected the abundance of insect food.

Food abundance for insectivorous birds (or birds that feed their chicks on insects) is generally greater in grassland systems, although there may be an issue of availability in the thick sward of habitats such as silage fields. In grassland systems, insect abundance will be affected by mowing regimes, stock type and density, and the medical treatment of the stock.

The important invertebrate groups in the diet of farmland birds include grasshoppers, spiders, leaf beetles, weevils, aphids, craneflies, and caterpillars of sawflies, butterflies and moths. There are indications that the declining bird species are more dependent on groups of larger insects, such as butterflies, moths, beetles and grasshoppers, which are showing some evidence of decline on farmland.

1.1.4 Soil invertebrates

The abundance and availability of soil invertebrates (especially earthworms and leatherjackets) are particularly affected by drainage, cultivation and the level and type of nutrient input. Soil invertebrate populations are generally greatest under permanent grassland, and lowest on arable farmland that is ploughed annually. Even when abundance is high, tall swards can reduce the ability of birds to probe into the ground and so access this food. Drainage of grasslands has reduced the ability for wading birds such as snipe to probe for soil invertebrates on farmland.

On grassland, the form of fertilisers and pesticides used can affect soil invertebrates. Soil invertebrates generally benefit from moderate applications of organic manures but decrease under high applications, particularly of slurry. The residues of some veterinary medicines have been shown to affect the survival, reproduction and development of invertebrates associated with the degradation of dung.

On arable land, soil invertebrate populations have declined with the loss of grass leys within the rotation, the use of inorganic fertilisers in place of farmyard manure, and the increased frequency of cultivations.

1.1.5 Habitat diversity

To a large degree, biodiversity is dependent on habitat diversity at every spatial scale.

- Mixed farms (or a mix of arable and pastoral farms within the landscape) will hold more species than all-arable or all-grass farms.
- Arable farms with a wide crop rotation will hold more species than monocultures of one crop type.
- Crops with a wide diversity of plants (weeds) will hold more associated species than clean crops.
- Grasslands with diverse sward structure, topography and water levels will hold more species than a uniform sward.
- Farms with a wide variety of hedges and field margins will hold more species than farms where all the hedges are the same height, or without field margin habitats.

In general, habitat diversity has decreased on farmland over the last 50 years due to the decline in mixed farming, simplification of arable rotations, improved weed control, grassland improvement and field boundary removal. The effects on individual bird species are that opportunities to forage for their food requirements around the year, or their ability to find both nesting and feeding habitats in close proximity, are more limited in a less diverse environment.

1.1.6 Protection from predation

A number of predators have increased during the period of declines of farmland birds (eg foxes, mustelids and crows) owing, at least in part, to the decline in the numbers of gamekeepers. Many birds of prey are also recovering from the declines resulting from the use of DDT and other organo-chlorine pesticides.

There is a widely held perception that increases in numbers of predators have caused the decline in farmland birds. In reality, any effects of predation are in addition to the impacts of the agricultural changes described in this chapter. In some cases, changes in land management have made some species more vulnerable to predation. These are described in the predation section (1.8).

It has been demonstrated experimentally that grey partridge breeding density increases with legal predator control. For this species, the Game Conservancy Trust views the ideal conservation package as including legal predator control in conjunction with land management to provide good quality nesting habitat, an abundance of insect food for the chicks and seed food for the adults.

Table 1.2 summarises the main changes in farming that have affected farmland bird populations. These broad changes, together with the effects of predation, are covered in more detail in the following sections. In addition, other factors such as more efficient harvesting machinery and the exclusion of birds from grain stores and animal feed stores have exacerbated the decline of food resources available to birds.

Table 1.2 Changes in farming since 1950 that have affected wildlife populations.

Change	Impact on wildlife
Decline in mixed farming	Reduced variety of habitats leading to loss of biodiversity
Agricultural improvement of grassland	Loss of unimproved grassland Switch from hay to silage causing earlier and more frequent cutting Less diversity of plants, insects and sward structure Loss of seed sources in grassland Increased stock trampling of nests
Changes in crops grown	Less spring tillage for nesting birds Fewer winter stubbles for seed food Earlier thicker growth reduces nesting and foraging opportunities Earlier harvesting damages nests Less crop diversity reduces biodiversity
Increase in pesticide use	Fewer flowering plants Fewer insects Polluted watercourses
Drainage	Loss of wet grasslands Loss of wet areas on farmland used for foraging
Hedgerow and field margin habitat removal	Loss of food, nest sites and shelter

Further reading

Aebischer N J, Green R E and Evans A D (2000) From science to recovery: four case studies of how research has translated into conservation action in the UK. Pp 173–181 in N J Aebischer, A D Evans, P V Grice and J A Vickery (eds) *Ecology and Conservation of Lowland Farmland Birds*. British Ornithologists' Union, Tring.

Anderson G Q A, Bradbury R B and Evans A D (2001) *Evidence for the effects of agricultural intensification on wild bird populations in the UK*. RSPB Research Report No 3.

Newton I (2004) The recent declines of farmland bird populations in Britain: an appraisal of causal factors and conservation actions. *Ibis* 146: 579–600.

Vickery J A, Bradbury R B, Henderson I G, Eaton M A and Grice P V (2004) The role of agri-environment schemes and farm management practices in reversing the decline of farmland birds in England. *Biological Conservation* 119: 19–39.

Wadsworth R A, Carey P D, Heard M S, Hill M O, Hinsley S A, Meek W R, Pannell D J, Ponder,V, Renwick A W and James K L (2003) A review of research into the environmental and socio-economic impacts of contemporary and alternative arable cropping systems. Report by the Centre for Ecology and Hydrology on *http://www.defra.gov.uk/environment/gm/research/pdf/epg_1-5-99.pdf*

Wilson J D, Morris A J, Arroyo B E, Clark S C and Bradbury R B (1999) A review of the abundance and diversity of invertebrate and plant foods of granivorous birds in northern Europe in relation to agricultural change. *Agriculture, Ecosystems and Environment* 70: 13–20.

1.2 Mixed farming

Mixed farms hold a greater diversity of habitats than specialised farming systems (Plate 1.4). This leads to a greater diversity of plants and animals and a greater range of opportunities for mobile species such as birds to explore. On mixed farms, the arable component generally provides more seed food for seed-eating birds and the grass component generally provides more insect and soil invertebrate food for insectivorous birds and species that feed their chicks on insects.

A mixture of arable and grassland formed the dominant farming system in the UK before the mechanisation of agriculture. After this, arable crops for horses were no longer required in pastoral systems, and artificial inputs replaced the need for grass leys in arable systems.

Diversity of species is largely the product of habitat diversity and, within agricultural systems, a mixed enterprise with both livestock and arable production offers more habitat diversity than more specialised enterprises. In terms of less mobile species, such as plants and invertebrates, mixed farms can cater for arable specialists and grassland specialists. The overall abundance and species-richness of insects is proportional to the number of patch types in the landscape. For more mobile groups, such as birds, there are more opportunities to benefit from the resources provided by both the grass and arable components of a mixed system.

In general, an arable element within a modern pastoral system will provide seeds in the form of arable weed seeds and seeds of crops, critical for a number of seed-eating birds. To an extent, seed food availability on grassland systems has declined, not only with the loss of an arable component, but also with changes in grassland management, such as use of herbicides and inorganic fertiliser, higher stocking rates, regular reseeding and the switch from hay to silage production. A

Plate 1.4 A mix of arable and livestock farming offers habitat for a greater range of species.

Andy Hay/rspb-images.com

grassland element within an arable enterprise can provide a greater abundance of accessible soil invertebrates, critical for some species such as the lapwing, starling and song thrush.

Some features found specifically on mixed farms offer additional resources. For example, undersown cereals provide a wealth of insects, particularly sawfly larvae for grey partridge chicks and summer food for corn buntings.

In some cases, the two enterprises complement each other in providing a continuous food resource through the year. For example, the linnet is dependent on small, oil-rich seeds throughout the year, which may be supplied by arable farmland in the summer, but early broods of linnets are often dependent on dandelion seeds found in grasslands. The juxtaposition of spring crops and grazed pasture provides ideal nesting habitat adjacent to ideal foraging and chick-rearing habitat for lapwings. Skylarks nesting in arable crops will also often make use of grassland in a mixed farming landscape when foraging for insects.

When a lowland farm landscape contains no arable land, a number of species are lost. These are mainly species dependent on seeds of crops and annual weeds such as the grey partridge, skylark, tree sparrow and the buntings. Their local extinctions in areas dominated by improved grassland are exacerbated by these marginal habitats holding lower densities in the first place. As little as 10% of the land in arable cropping could be sufficient to maintain populations of these species in lowland grassland landscapes.

Insectivorous species and seed eaters that are not dependent on annual weeds, but also feed on perennial seeds of grasslands and other habitats, such as the chaffinch, greenfinch and goldfinch, are largely unaffected by the loss of arable farming in pastoral areas. One exception is the whitethroat. This may be linked to its use of rough grass margins and ditch banks in arable fields. In grassland systems, such habitat is often grazed by stock, removing opportunities for nesting whitethroats.

Arable pockets in grassland systems may occur where use is made of arable fodder crops such as undersown cereals, whole-crop cereals and brassica fodder crops. Whole-crop cereals may be less beneficial in terms of spilt grain as they are harvested before the grain is ripened, but the early harvest may encourage a greater flush of weed growth. In pastoral areas, where the amount of arable cultivation is never going to be high, the key features from a bird conservation perspective are spring-sowing, stubble retention and rough grass margins.

Further reading

Evans A D (1997) The importance of mixed farming for seed-eating birds in the UK. In D J Pain and M W Pienkowski (eds) *Farming and Birds in Europe: The Common Agricultural Policy and its implications for bird conservation*. Academic Press, London.

Robinson R A, Wilson J D and Crick H Q P (2001) The importance of arable habitat for farmland birds in grassland landscapes. *Journal of Applied Ecology* 38: 1059–1069.

1.3 Pesticides

The main effects of pesticides on bird populations today are their impact on the abundance of seed and insect food on farmland. Pesticide use has been proven to be a cause of decline of the grey partridge, and there is strong evidence that it has also caused declines in linnets, yellowhammers and corn buntings. There is also some evidence of links with the declines of other species.

The effects of pesticides on birds can be direct toxic effects (such as the impact of organo-chlorine pesticides on birds of prey in the 1950s and 1960s) or indirect, through the loss of food sources in terms of plants or insects removed by the chemicals. The use of pesticides also has direct implications for the conservation of plants and invertebrates on farmland, and may have indirect effects on other herbivorous and insectivorous vertebrates, such as hares and bats. The approval process for existing and new pesticides has now largely removed the direct effects on birds. However, there are issues with certain products, such as rodenticides, that are minimised by careful use. Direct toxic effects are not considered further in this section.

The bulk of the declining farmland birds feed on seeds, and most of these feed their chicks on insects. The leaves and seeds of many arable weeds are important food for farmland birds, and many are also hosts to important invertebrate food items. The three main ways in which pesticides can reduce the food supply for these birds are:

- the direct removal of invertebrates by insecticides
- the removal of the food plants of insects by herbicides
- the removal of plants that provide food for birds by herbicides.

Many fungicides, which are used more frequently than insecticides, also have insecticidal properties, but the importance of these is still not fully understood. During research on yellowhammers, fungicides were found to have some impact on key invertebrate food items, but no significant effect on yellowhammers was found.

Weeds of arable land are probably the most seriously declining plant group in the British Isles. The most likely cause of their decline is the increase in the use of herbicides and artificial fertilisers. Tillage methods and crop density may also affect seed viability and the scale of competition for resources that they face.

The impact of insecticides on non-target invertebrates is variable. Pirimicarb (a carbamate) and tau-fluvalinate (a pyrethroid) affect significantly fewer non-target insect groups than other pyrethroids. The pyrethroids are variable in the spectrum of invertebrate groups affected, but generally do less damage than the organophosphates, such as dimethoate. Scientific monitoring shows that summer insecticides (applied after March) have a more damaging effect on insect food available to birds in the breeding season than autumn/winter insecticides. This is probably because most of the important insect groups are not in the cropped area before the spring.

There is also a link between herbicide use and the abundance of insects. Cereals that do not receive an autumn herbicide are significantly weedier and hold higher populations of invertebrates, particularly beetles and bugs. Herbicide-free crops will not be acceptable to many farmers, but selective sprays to take out the grass weeds and cleavers but leave the other broad-leaved weeds and associated insects in the headlands are a possibility. This practice, known as conservation headlands, was developed by the Game Conservancy Trust and has been proven to benefit chick survival of grey partridges. Other projects are currently investigating the selective use of herbicides that control pernicious weeds in the field centres while leaving species that have less impact on crop yields and which are valuable sources of food for birds, such as groundsel.

The sharp decline of many seed-eating birds on farmland occurred between the mid-1970s and the mid-1980s. The use of herbicides, fungicides, insecticides and molluscicides on cereal crops all increased during this period (Plate 1.5). However, increased pesticide use is not the main cause of decline in all cases. This period coincided with other important changes in agriculture.

Work on the grey partridge by the Game Conservancy Trust pioneered the research on the indirect effects of pesticides on birds. The research in Sussex was able to show that the survival rate of grey partridge chicks was dependent on the proportion of the diet made up of insects, and this was directly linked to the relative abundance of insects in the surrounding cereal fields. Many of the important insect groups taken by grey partridge chicks are those which feed on arable weeds and are effectively removed by efficient herbicide regimes.

The Game Conservancy Trust also found strong correlations between the use of herbicides and insecticides, and the occurrence of weeds, abundance of key insect groups for birds and bird densities between 1970 and 1995 on their study site in West Sussex. Herbicide use was already widespread in 1970, but the number of applications per crop increased throughout the study period. There were also changes in the types of herbicide used and the efficacy of herbicides (measured by the number of weed groups susceptible to chemical control) increased. Fungicide and insecticide use became widespread, and the average number of applications per crop and the number of active ingredients used increased during this time. Insecticide use affected all of the key insect groups studied, but was most significant when applied in the spring or summer. All groups were also affected by insecticide use in the previous year, indicating long-term consequences of insecticide applications. Spring herbicides generally had a greater effect on broad-leaved weed numbers than autumn applications.

Research on corn buntings in the same Sussex study area showed a similar dependence on insect food for the chicks. Corn buntings selectively foraged, and reared nestlings with better body condition, in areas with a greater abundance of the key insect groups. These areas received significantly fewer pesticides (herbicides, fungicides and insecticides) than areas not used for foraging.

Andy Hay/rspb-images.com

Plate 1.5
The average number of pesticides applied to crops increased through the period of the steepest farmland bird declines, but pesticides are not responsible for all of the declines.

Fields receiving insecticides early in the nesting season, when yellowhammer chicks are fed exclusively on invertebrates, were used less by foraging yellowhammers, and the body condition of yellowhammer nestlings in adjacent field boundaries was reduced. Later in the nesting season, the diet of yellowhammer chicks can be supplemented by ripening cereal grains, although insects remain important.

Evidence for effects on skylarks are equivocal, with some European studies suggesting a demonstrable effect and others suggesting no effect at all. At present, the best available evidence suggests that while pesticides can have a negative impact on populations, these impacts are small when compared to those of crop structure. Currently, factors other than pesticide use are more likely to be limiting populations. Further research into the impact of summer use of insecticides on chick condition and survival are ongoing in the UK.

The decline of the tree sparrow may be linked to either the decline in winter seed food or the decline of summer insect food, both of which may be linked to pesticide use. Since their decline, tree sparrow populations are often associated with watercourses or large water bodies. In a study at Rutland Water, first broods and some second broods were largely reared on aquatic insects in some years, probably when insect availability on farmland was in short supply.

The declines of linnet and turtle dove are most likely linked to the direct effect of herbicides removing seed food sources on arable land. These seed-eating birds do not rely on insect food for their chicks. The diet of these species has changed from predominantly weed seeds in the 1960s to oilseed rape (and cereal grain in the case of turtle doves) in the 1990s. Whilst there is no shortage of these crops in many areas, availability of the seed is seasonal and so there will be long periods of food shortage.

The dependence of many species on seed food through the winter on farmland means that seed-rich areas such as overwinter stubbles are very important for overwinter survival. However, the efficiency of herbicides in removing weeds, together with the efficiency of modern combine harvesters in removing the crop means that many overwinter stubbles lack the food resource required and weedy stubbles are far more productive for birds than clean stubbles. New research suggests that clean stubbles are largely avoided by granivorous passerines, with large, overwintering flocks being concentrated on a small number of weedier stubbles. Before herbicides were so effective, it is likely that seed-bearing plants were found more generally on farmland, and that overwinter stubbles in the landscape now have to fill the role of a greater diversity of feeding opportunities that existed in the past. There are also additional factors such as fewer sites where birds can make use of food for out-wintered stock and exclusion from grain stores.

In grassland systems, herbicides are generally restricted to the control of perennial weeds and the establishment of new grass leys, and insecticide use is rare. Overall, the effects of pesticides in reducing food resources for birds has been small in comparison to the effects of other changes in grassland management.

The residues of anthelmintics such as avermectins in the dung of livestock following worming treatment have been shown to affect the survival, reproduction and development of invertebrates associated with the degradation of dung. As well as being important in their own right, these invertebrates can be important food for bats and birds such as pipits, wagtails, waders, thrushes, swallows and crows. Invertebrates associated with dung are especially important for the chough during the breeding season.

Further reading

Brickle N W, Harper D G C, Aebischer N J and Cockayne S J (2000) Effects of agricultural intensification on the breeding success of corn buntings *Miliaria calandra*. *Journal of Applied Ecology* 37: 742–755.

Campbell L H, Avery M I, Donald P, Evans A D, Green R E and Wilson J D (1997). *A review of the indirect effects of pesticides on birds*. JNCC Report No 227. Joint Nature Conservation Committee, Peterborough, UK.

Ewald J A and Aebischer N J (1999) *Pesticide use, avian food resources and bird densities in Sussex*. JNCC Report No 296. Joint Nature Conservation Committee, Peterborough, UK.

Morris A J, Wilson J D, Whittingham M J and Bradbury R B (in press) Indirect effects of pesticides on breeding yellowhammers *Emberiza citrinella*. *Agriculture, Ecosystems and Environment*.

Potts G R (1986) *The Partridge: Pesticides, Predation and Conservation*. Collins, London.

Rands M R W and Sotherton N W (1986) Pesticide use on cereal crops and changes in the abundance of butterflies on arable farmland in England. *Biological Conservation* 36: 71–82.

1.4 Crops

The main changes in arable cropping since the 1970s has been the increase in block cropping, the replacement of spring-sown crops with winter-sown crops and the introduction of set-aside.

Large areas of one crop type limit the habitat available to less mobile species, such as invertebrates, and reduce the foraging opportunities for nesting birds when they need to feed in close proximity to the nest. The loss of spring cropping has reduced the availability of nesting habitat for some ground-nesting birds, and reduced the amount of overwinter stubbles available for seed-eating birds in the early winter. The introduction of set-aside may have had some benefits for some species especially in the winter, but a small fraction provides summer benefits due to the timing of weed control.

1.4.1 Changes in crop types and rotations

Arable cropping in the UK has always been dominated by cereals, but broad-leaved crops are used in rotations to act as break crops to help with the control of weeds, pests and diseases. Cereals made up over 80% of the arable area of Great Britain in 1974, and this declined to about 70% in 2000. A similar change has occurred in Northern Ireland (Table 1.3). Areas of some crops such as sugar beet, legumes and potatoes have remained fairly static. Crops that have increased in area include wheat and oilseed rape, and those that have declined are other cereals and root fodder crops (especially in Scotland). The loss of root fodder crops is of particular importance for farmland bird conservation, as it is an important source of weed seeds through the winter.

However, the three main changes have been the scale at which blocks of individual crops are grown, the decline in the area of spring sowing of cereals and the introduction of set-aside. Mechanisation has led to a dramatic increase in average arable field size. In addition, having fields with the same crop in one block reduces the time needed during management operations. In England and Wales, the proportion of cereals sown in the spring declined from about 55% in 1974 to less than 10% in 2000. In Scotland, the proportion declined from 93% to less than 60% over the same period, and in Northern Ireland it declined from 68% to 42% between 1981 and 2004. Obligatory set-aside was introduced in 1992 after the main period of decline of farmland bird populations.

Table 1.3 Changes to arable cropping in the UK. Figures given are the percentage of the arable area.

	England and Wales		Scotland		Northern Ireland	
Crop	**1974**	**2000**	**1974**	**2000**	**1981**	**2004**
Wheat	31	47	6	18	1	16
Winter barley	6	12	–	10	4	8
Spring barley	41	6	63	43	68	42
Oats	4	2	13	4	4	5
Other cereals	1	1	–	–	1	–
Oilseed rape	1	7	2	6	–	1
Other brassicas	2	–	1	–	–	–
Peas/legumes	2	2	–	–	–	–
Linseed	–	2	–	–	–	–
Potatoes	4	3	6	5	17	11
Field beans	2	3	–	–	–	–
Sugar beet	5	4	–	–	–	–
Root crops	1	–	10	–	–	–
Set-aside	–	11	–	13	–	6
Other	–	–	–	–	4	12

Andy Hay/rspb-images.com

Plate 1.6
Sowing large areas with one crop type reduces the diversity of habitats for wildlife and the range of opportunities for nesting and foraging birds.

1.4.2 Block cropping

If large areas are sown with a single crop type then the effect on the range of insects in the crop is exacerbated by the distance to the nearest refuge where these insects can survive until the field is next under a suitable crop type. For less mobile species, block cropping (see Plate 1.6) can cause irrecoverable declines.

For birds too, large areas under one crop type can limit the foraging opportunities, especially during the nesting season, when they need to feed within close proximity of the nest. Cereal crops provide good habitat for a range of farmland birds, but a wider range of crops will provide more diversity and therefore create opportunities for a greater range of plants and invertebrates to thrive.

Different crops provide different resources. Cereals provide grain food for buntings, a suitable sward for nesting skylarks and brood-rearing habitat for partridges. Oilseed rape provides seed food for linnet chicks, and may be the cause of the recovery of the population in some areas where weed seeds have effectively been removed by herbicide use. Broad-leaved crops in general support more broad-leaved weeds, although these are intensively controlled in some high value crops, such as potatoes and sugar beet.

1.4.3 The switch from spring to autumn sowing

The most dramatic change in crops grown, in terms of the impact on birds, is the switch from predominantly spring-sown to autumn-sown cereals. Crops sown in spring (Plate 1.7) provide bare tillage and sparse swards in the nesting season. These crops are harvested later which can be crucial for late nesting species such as corn buntings. They also enable the retention of stubbles through the winter that can provide seed food for birds.

Lapwings and stone-curlews are dependent on open spring crops for nesting in arable systems. The denser, taller vegetation of an autumn-sown crop will also affect access in the spring for other ground-nesting and foraging birds. The main factor limiting breeding success of the skylark was identified as crop height and density through the breeding season. As a result, spring cereals were more productive than winter cereals. Winter cereal swards opened up by leaving undrilled patches (known as skylark plots) boosted skylark breeding success by almost 50%.

Historically, post-harvest stubbles were more abundant before autumn-sowing led

R Winspear (The RSPB)

Plate 1.7
Spring cereals (on the left) provide benefits for wildlife, particularly ground-nesting birds.

to earlier cultivations. These stubbles were rarely left throughout the winter, but formed an important food source for birds in early winter, before other sources became available. Overwinter stubbles were not the only seed source for birds in the mid-20th century. Other seed sources have largely been removed by a combination of factors, including increased use and efficiency of herbicides. Therefore, winter stubbles can be a very important seed source in modern arable systems, although clean stubbles (without weeds) may be of very little value. In general, modern stubbles contain less seed food than stubbles in the past due to more efficient weed control in the preceding crop and more efficient harvesting machinery spilling less grain.

1.4.4 Set-aside

Set-aside was introduced on arable land as a production limitation measure, and the management rules when it was first introduced were particularly damaging for wildlife (eg cutting during the breeding season). Since then, the rules have been amended to protect most wildlife interests, but positive management for wildlife is still only undertaken on a relatively small proportion of the set-aside area (Table 1.4).

Natural regeneration is usually cereal stubble retained after harvest and allowed to develop a green cover through the winter. It can provide important winter seed food, although this varies, depending on the level of weed control in the preceding cereal crop and the amount of grain spilt during harvest. It is common practice to spray the green cover off with a total herbicide in early spring (after 15 April) to control grass weeds. The wildlife value of the set-aside during the breeding season is dependent on the timing of management. It is very limited after the herbicide application as nesting cover and insect abundance declines within a few weeks.

Table 1.4 Categories of set-aside management in Britain in 2000.

Set-aside management type	% set-aside area in Britain
Natural regeneration	65%
Industrial oilseed rape	14%
Other crops	12%
Grass cover	7%
Wild bird cover	2%

Industrial crops grown on set-aside generally have no greater value than other arable crops. Grass cover is generally created on fixed set-aside on poor yielding areas and field margins. These areas can build up large invertebrate populations and are often used to buffer watercourses, and this is generally a good option for wildlife. Finally, wild bird cover provides cover for game birds, and if seed-bearing crops are established, then it also provides seed food for birds through the winter.

In summary, wildlife benefits from the grass and wild bird cover options, and there are variable benefits from the natural regeneration set-aside.

Further reading

Buckingham D L, Evans A D, Morris T J, Orsman C J and Yaxley R (1999) Use of set-aside in winter by declining farmland bird species in the UK. *Bird Study* 46: 157–169.

Evans A D (1997) Seed-eaters, stubble fields and set-aside. *Proceedings, 1997 Brighton Crop Protection Conference*, pp 907–922. British Crop Protection Council, Farnham.

Hancock M H and Wilson J D (2003) Winter habitat associations of seed-eating passerines on Scottish farmland. *Bird Study* 50: 116–130.

Moorcroft D, Whittingham M J, Bradbury R B and Wilson J D (2002.) Stubble field prescriptions for granivorous birds – the role of vegetation cover and food abundance. *Journal of Applied Ecology* 39: 535–547.

Moorcroft D and Wilson J D (2000) The ecology of linnets on lowland farmland. Pp 173–181 in N J Aebischer, A D Evans, P V Grice and J A Vickery (eds) *Ecology and Conservation of Lowland Farmland Birds*. British Ornithologists' Union, Tring.

1.5 Grassland management

> The most significant effects of the changes seen in grassland management have been reductions in the availability of seed and invertebrate prey within the sward and the quality of grassland as a nesting habitat.

Many of the farmland birds that have declined have also undergone range contractions in the pastoral areas of the UK. Changes in grassland management have reduced suitable nesting habitat and made nests more vulnerable to machinery operations. Important food resources, such as seeds and invertebrates, have become less abundant. The loss of arable cultivations in pastoral areas has reduced the availability of weed seeds and cereal grains.

Birds use grassland for nesting, feeding, shelter and cover from predators. As requirements differ considerably between species, so responses to changing grassland management have varied. While some species may have benefited, others have suffered a deterioration of their preferred habitats. The links between changes in grassland management and changes in bird populations are not as well understood as those in arable systems. There is good evidence that the onset of mechanisation and the switch from hay to silage has caused the decline of the corncrake. For some species, potential mechanisms underlying declines can be inferred from the known effects of grassland intensification on their key requirements. Studies of yellowhammers, skylarks and corn buntings have shown these birds to actively avoid intensively managed pastures and meadows when selecting breeding territories and foraging habitat. Species like the wheatear and whinchat were once widely distributed on lowland grasslands, but are now almost entirely restricted to upland areas.

1.5.1 Changes in grassland management

Historically, the diversity of plants in grassland was determined mostly by geology, climate, hydrology, topography and management. As a result, there was a wide variety of grassland communities. In modern improved swards, management is the overriding factor influencing the grassland community.

By the eighteenth century, efforts were being made to improve the agricultural productivity of the UK's grasslands through sowing desired mixtures of grass seed and the introduction of legumes. However, it was after the Second World War when the most dramatic changes were seen. This was brought about through advances in plant breeding, grassland management knowledge, increased artificial inputs, reseeding, mechanisation and drainage. These changes were accelerated by government incentives to boost agricultural productivity and the provision of an advisory body to disseminate information on how this could be achieved. By the end of the twentieth century, an estimated 97% of unimproved lowland grassland had been lost. The Countryside Survey 2000 noted continued loss of the surviving parcels of less improved grasslands in Scotland and Northern Ireland between 1990 and 1998. England and Wales experienced deterioration in the quality of the remaining unimproved grassland through increasing levels of fertility. The changes in grassland management and their effects on birds are discussed below and in the section on field drainage (1.7).

Fertiliser

Use of inorganic nitrogen fertiliser on grassland increased two- to three-fold between the 1940s and the late 1980s, and by the late 1990s, 86% of all UK grasslands received inputs of inorganic fertilisers. Artificial fertilisers caused a marked reduction in botanical diversity. Many wild plants have evolved in ecosystems that are nutrient poor. Consequently, they cannot tolerate high concentrations of nutrients or are out-competed by species able to respond to increased fertility. Grasses, particularly agricultural varieties that have been bred to respond quickly to high nutrient levels, monopolize space and light and become dominant. A few broad-leaved plants that thrive in high nutrient soils such as docks and nettles are also strongly competitive as weeds. Herbs and less competitive grasses begin to disappear from botanically diverse grasslands at the lowest levels of artificial nitrogen application (25 Kg N/ha/year). Applications of phosphorous also severely affect plant diversity.

Slurry has replaced solid farmyard manure as the main organic fertiliser applied to grassland. Slurry releases its nutrients faster than solid farmyard manure, to the detriment of botanical diversity.

Grazing

The agricultural improvements of grasslands and subsidies based on stock numbers led to an increase in livestock densities, particularly sheep, in pastoral areas. Between 1976 and 1997, the total number of sheep in the UK increased by over 50%. Coupled with more efficient grazing systems, the extent of short even swards with little structural diversity has increased. Simultaneously, some isolated and nutrient-poor pastures have become uneconomic to graze, creating a sward too rank for many grassland nesting birds.

The switch from hay to silage

Silage has replaced hay as the dominant winter fodder. It has been a revolutionary change for livestock farmers, particularly in areas where weather made good quality forage difficult to make. A barn full of poor quality hay, when alternative feeds were far less easily obtained, had consequences for the welfare of livestock over the winter.

The switch from hay to silage has facilitated many changes in grassland management. Dense, heavily fertilised ryegrass swards that would be difficult to turn into hay can be conserved as silage. A greater emphasis on optimising yields has resulted in earlier and repeated cuts through the summer. These factors have combined to make intensively managed silage fields generally poor nesting and foraging habitats (Plate 1.8).

Andy Hay/rspb-images.com

Plate 1.8 Intensively managed silage fields are generally very poor for ground-nesting birds due to the timing and frequency of cutting.

Increasing mechanisation

The scale and frequency of mechanical operations such as muck spreading, harrowing, rolling and topping has increased significantly. The speed and timing of the grass harvest has been particularly altered by mechanisation. In the past, harvest was a drawn out task with a long time lag between cutting the first and last fields. Modern machinery has allowed this task to be completed in a much shorter time.

Reseeding

The desire to maintain sown agricultural grasses and legumes as the main constituents of agricultural grasslands has increased the extent of reseeding. Many permanent pastures have been replaced by short-term leys. Historically, reseeding involved the sowing of relatively complex seed mixes containing many grasses, legumes and forage herb species. Modern mixes are usually simple and are generally purchased according to the intended duration and management of the sward. Short-term leys of a few years typically contain high yielding Italian or hybrid ryegrass, while mixes intended for longer term leys are based on perennial ryegrass. Regular reseeding with simple seed mixes has significantly reduced the diversity of plants found in agricultural grasslands.

1.5.2 Effects on birds

Nesting habitat

Dense, fast growing ryegrass swards are poor nesting habitats for birds. Birds have strong preferences for particular grassland structures. The lapwing prefers uneven swards that camouflage the nest but are short enough for the sitting bird to look out for predators, while the snipe requires taller vegetation to conceal its nest near short grassland where it can probe for food. Many modern swards lack the structural diversity required by ground-nesting birds. Grazing can be manipulated to maintain the sward structure at a desired level, but the stocking levels required to achieve this increase with grassland improvement. Higher stocking rates increase the chance of nest trampling, particularly for waders. Though breeding waders can lay a second clutch of eggs following trampling, some birds will not re-lay, especially later in the summer.

The switch from hay to silage and the speed and timing of the modern grass harvest has reduced nesting habitat for species such as the corncrake, curlew, skylark, quail, whinchat and corn bunting. Silage fields are poor nesting habitats due to their dense fast growth, and early and frequent cutting. Nesting attempts are often destroyed by mowing, but can be avoided if their timing is right. When the grass harvest was extended over a number of months, it provided several nesting opportunities through the breeding season for birds like the skylark and afforded later ground-nesting birds, such as curlew, the opportunity to breed. Skylarks rear several broods through the summer, but they need at least seven weeks between cuts to be successful. The loss of nests and chicks of corncrakes increased with faster mowing and through machinery working from the outside, creating open ground that corncrakes are reticent to cross. Nests of species such as lapwing and skylark are also lost to mechanical operations such as muck spreading (Plate 1.9), harrowing, rolling and topping during the breeding season.

Invertebrate food

Invertebrates are an important component of the diet of many farmland birds, especially during the summer. Some species are completely insectivorous all year round, but others require insects for their chicks. Birds such as starlings and thrushes take invertebrates from the soil, while buntings, finches and larks, take invertebrates from within the vegetation,

Andy Hay/rspb-images.com

Plate 1.9 Operations such as muck spreading can destroy nests during the breeding season.

and swallows and flycatchers hawk for insects above the sward.

Some soil invertebrates, such as earthworms, may benefit from the higher fertility levels found in modern grasslands, but heavy use of slurry and regular cultivations associated with short-term leys cause a decline in numbers. Earthworms decline after cultivations, and can take up to four years to recover to levels found in permanent pasture. Soil invertebrate feeders such as lapwing, golden plover, starling, redwing and fieldfare favour permanent grassland in the winter.

Invertebrates within the vegetation have been particularly affected by changes in grassland management. Frequent cutting and lack of features such as dead grass at the base of the sward and taller tussocks, reduce opportunities for invertebrates to complete their lifecycles, especially larger invertebrates like grasshoppers and moth larvae. Loss of botanical diversity has reduced the variety of invertebrate food plants in the sward. Many invertebrates require particular plant species. For example, some butterflies feed on a variety of nectar plants but their caterpillars feed on one specific species. Other invertebrate food resources such as nectar, pollen and seed heads are also reduced in modern swards.

Like soil invertebrates, those found in the vegetation are also greatly disrupted by cultivations associated with reseeding. Some of the smaller, more mobile invertebrates can re-colonise the new sward quite quickly, while larger, longer-lived invertebrates take several years to return, if at all.

Fertiliser applications can cause many invertebrate groups such as beetles, flies, grasshoppers, crickets, springtails and centipedes to decline in abundance. Fertiliser applications are particularly damaging to larger invertebrate species. Many rare and declining birds, such as cirl bunting, corn bunting and lapwing, specialise in foraging for larger invertebrates. The average body mass of insects decreases with increased nitrogen applications and falls steeply once annual applications rise above 50 kg/ha.

Seed food

The loss of plant diversity caused by artificial fertilisers and reseeding reduces the variety and timing of seed production within a sward. As different plants set seed at different times of the year, a wider variety extends the period of seed provision. Heavy grazing also reduces seed production in the sward, and hence the opportunities for seed-eating birds such as finches and buntings.

Hay fields containing a variety of grasses and broad-leaved plants, many of which will have set seed before cutting, provide far more food for seed-eaters than silage fields managed as ryegrass monocultures, cut before seed is set. Less seed food is available through the winter too, when the resulting fodder is fed to livestock. The large quantities of seeds in dung from stock fed on seed-rich herbage is another important food source that is lost to these birds.

Other food sources

High stocking densities and efficient grazing systems generally prevent the build up of grass litter at the base of the sward reducing habitat for small mammals and consequently hunting kestrels and barn owls.

Wintering geese and swans favour fertilised ryegrass swards for the higher protein content in the forage. However, these species tend to be localised in their winter distribution and utilise a very small proportion of the grassland in the UK.

Further reading

Atkinson P W, Buckingham D L and Morris A J (2004) What factors determine where invertebrate-feeding birds forage in dry agricultural grasslands. In J A Vickery and A D Evans (eds) Ecology and Conservation of Lowland Farmland Birds II: The Road to Recovery. *Ibis* 146 (suppl 2): 99–107.

Hopkins A (ed) (2001) *Grass, its production & utilization, 3rd Edition.* Blackwell Science Ltd, Oxford.

Morris M G (2000) The effects of structure and its dynamics on the ecology and conservation of arthropods in British grasslands. *Biological Conservation* 95: 129–124.

O'Conner R J and Shrubb M (1996) *Farming and birds.* Cambridge University Press, Cambridge, UK.

Shrubb M (2004) *Birds, scythes and combines.* Cambridge University Press, Cambridge, UK.

Vickery J A, Tallowin J R., Feber R E, Asteraki E J, Atkinson P W, Fuller R J and Brown V K (2001) The management of lowland neutral grasslands in Britain: effects of agricultural practices on birds and their food resources. *Journal of Applied Ecology* 38: 647–664.

1.6 Field boundaries and margins

Since 1945, field sizes have generally increased, especially on arable land, with a resulting loss of field boundary and field margin habitats, which can be a valuable source of nesting habitat, seed food and insect food for some species. A number of in-field nesting species avoid tall field boundary features, and may have benefited from increasing field size, but loss of field margin habitats may still have been detrimental to them by reducing in-field insect populations.

The length of hedgerows in the UK has declined since 1945. This corresponded with increasing field size since 1945, especially on arable land, and the loss of field margin habitats, which are important for boosting invertebrate abundance in many farming systems. The recent increase in hedgerow length since 1990 is largely due to restoration of relict hedges, rather than reversing the trend for increasing field size.

Although many of the declining farmland birds nest in hedgerows, loss of nesting habitat has not been identified as a main factor in the decline of any of these species, although it can determine nesting densities in some areas. However, loss of invertebrate food for chicks has been implicated in the declines of several species including the grey partridge and corn bunting. Hedgerows and field margin habitats are a rich source of insect life, both in terms of potential foraging areas directly, and also as overwintering habitat for many insects that move into arable crops in the spring. Although the increased use of pesticides is known to have reduced the availability of insect food for some species, loss of field boundaries and field margin habitats is probably another important factor. Whitethroats, tree sparrows, yellowhammers and corn buntings are known to preferentially use rough grass margins of arable fields to forage for insect food for their chicks. Yellowhammer and whitethroat nesting densities along a hedgerow are correlated to the width of the adjacent field margin to provide nesting habitat and insect food for the chicks (Plates 1.10 and 1.11). Likewise, nesting corn buntings have significantly more rough grass field margins within 250 m of the nest site than would be expected by a random distribution.

The quality of the field boundary and margin habitats is also key to their importance for wildlife. A simple hedged boundary that is cropped up to the edge (a practice encouraged under past subsidy payments based on cropped area and numbers of stock) is far less of a wildlife resource than a hedge with an uncropped margin. Species differ in their preference for field boundary type. Many species that nest in hedgerows prefer taller hedges with an abundance of hedgerow trees. However, whitethroats, linnets and yellowhammers tend to prefer shorter hedges with few trees. Well-maintained hedgerows also provide cover from the

Chris Gomersall/rspb-images.com

Gordon Langsbury/rspb-images.com

Plates 1.10 and 1.11 The numbers of yellowhammers (left) and whitethroats (right) nesting in a hedge relates to the width of the adjacent field margin.

elements and predators. It may be that intensive flailing of hedgerows provides less effective nesting cover and increases the risk of predation.

The bulk of the wildlife in most fields will be found within the field boundary, the uncropped margin and the crop edge. The rare exceptions will be some unimproved or semi-improved grasslands. The wildlife interest of the field boundaries and margins is determined by:

- the area of the uncropped land and the degree to which it forms a network around the farm – in general farms with smaller average field size will hold more wildlife than farms with larger fields, and those with wide margins are better than farms with fields cropped or grazed right up to the field boundary
- the type and management of the boundary and uncropped areas – 'living boundaries', such as hedges and ditches, are a rich wildlife resource; grass banks and dry-stone walls can harbour some wildlife interest; fences and featureless boundaries have no wildlife interest *per se*. All boundary types can be valuable for wildlife if appropriate management is undertaken.

Many research studies have highlighted the importance of field margin type and management in determining the densities of farmland birds.

The grey partridge, turtle dove, thrushes, sparrows, finches and yellowhammer are all species largely dependent on field boundaries and uncropped habitats for nesting. Although most of these feed within the fields for seeds or invertebrates, the field boundaries and margins can also be a rich food resource for them. Hedgerows and ditches can provide a rich source of insect food, although some farmland birds prefer to forage in more open habitats. Rough grass field margins provide overwintering habitat for a wide range of insects and spiders associated with farmland. Wild flower margins can also provide wintering habitat, but also provide a pollen and nectar source that can further boost insect abundance. The crop edge generally supports more weeds than the centre of arable fields and can therefore be a source of seed food. Cultivated, uncropped margins or conservation headlands (crop edges receiving no summer insecticides or broad-spectrum herbicides) can establish a rich arable flora, particularly on lighter soils, that provides seed food for birds and food plants for a wide range of insects.

In contrast, some species, such as breeding waders and skylarks, avoid areas with tall field boundary features to evade predators. Areas supporting important numbers of these species are best maintained as open habitats with appropriate boundaries such as grass strips. Grass banks or strips provide a rich invertebrate resource without compromising the landscape character for open field species. Indeed, the capacity of these features for holding overwintering invertebrates that then move up to 250 m into the crop in the spring should improve the foraging potential for these species.

Further reading

Morris A J, Whittingham M J, Bradbury R B, Wilson J D, Kyrkos A, Buckingham D L and Evans A D (2001) Foraging habitat selection by yellowhammers *Emberiza citrinella* nesting in agriculturally contrasting regions in lowland England. *Biological Conservation* 101: 197–210.

Perkins A J, Whittingham M J, Morris A J and Bradbury R B (2002) Use of field margins by foraging yellowhammers *Emberiza citrinella*. *Agriculture, Ecosystems & Environment* 93: 413–420.

Stoate C and Szczur J (2001) Whitethroat *Sylvia communis* and Yellowhammer *Emberiza citrinella* nesting success and breeding distribution in relation to field boundary vegetation. *Bird Study* 48: 229–235.

Vickery J, Carter N and Fuller R J (2002) The potential value of managed cereal field margins as foraging habitats for farmland birds in the UK. *Agriculture, Ecosystems and Environment* 89: 41–52.

1.7 Field drainage

The most obvious consequence of land drainage on wildlife has been the loss of wildlife-rich wet grassland habitats, but there has also been a reduction in general wet features on all farm types that has been detrimental to more widespread farmland species.

Land drainage has a long history in the UK, going back to Roman times, but became widespread after the enclosure acts of the 18th century. About half of the agricultural land in the UK is dependent on artificial drainage. Wet grassland habitats such as coastal grazing marsh, floodplain grassland, wash-lands and water meadows were valued agriculturally for their early spring growth, resistance to drought and the influx of fertile sediment. However, agricultural improvement of these areas through better land drainage, grassland improvement and conversion to arable resulted in the loss of most wet grassland with a high conservation value. Wet grassland habitats held a great diversity of wildlife, such as aquatic plants and invertebrates, and waders such as the lapwing, redshank and snipe (Plate 1.12).

Recent research on farmland birds has also indicated the importance of wet features on farmland for species such as the yellow wagtail, song thrush, tree sparrow and reed bunting. Small wet features are important to these species on all farm types when they are feeding their chicks on invertebrates.

Another effect of this activity on birds and other wildlife has been the straightening and lowering of riverbeds to achieve adequate freeboard for discharge and reduce flooding. This has caused an enormous reduction in a variety of natural river system habitats such as shoals, beaches, islands, riffles, pools and oxbow lakes.

Restoration of wet grassland habitats is a priority conservation action in suitable target areas, but there is scope for introducing small wet features onto farmland more widely through the creation of accessible ponds and scrapes, and the re-profiling and raising of water levels in ditches.

Further reading

Wilson A M, Ausden M and Milsom T P (2004) Changes in breeding wader populations on lowland wet grasslands in England and Wales: causes and potential solutions. *Ibis* 146 (suppl 2): 32–40.

Andy Hay/rspb-images.com

Plate 1.12 Snipe numbers have declined with the loss of wet grasslands, but a much wider range of wildlife has been affected by the general loss of wet features and damp areas on farmland.

1.8 Predation

Predation can be an additive factor if it prevents a population from recovering the losses caused by agricultural intensification, or if habitat changes make a species more vulnerable to predation. Habitat change can increase predation rates if populations become concentrated into smaller areas, species are forced to nest or feed in more vulnerable situations, or if the behaviour of birds is affected (eg chicks become more vocal or more mobile due to the shortage of food).

In general, if birds are able to nest successfully, find plenty of food for their chicks and survive through the winter, then the population is able to withstand predation pressure. This is illustrated by the wide range of common prey species that have stable population trends. Research has shown that all of the declining farmland bird species are failing at some stage of the life cycle due to one or more of the agricultural changes described in earlier sections. Increased predation can be an additional factor if it prevents a population from recovering the losses caused by agricultural intensification, or if habitat changes increase the species' vulnerability to predation.

The one species of lowland farmland for which the impacts of predation has been proven is the grey partridge. Research has shown that, although insect food availability for chicks is the main factor driving the decline, predation of eggs, chicks and incubating hens can exacerbate the problem. Research studies in Wiltshire and Sussex demonstrated a two- to three-fold increase in grey partridge breeding densities with legal predator control. The Game Conservancy Trust believes that predator control in conjunction with positive land management is required to restore populations of this species.

For other species, predation may have an additional effect if habitat changes make them more vulnerable to predation through one of the following means.

- If habitat availability for a prey species is reduced so that the prey are concentrated into small areas of the farm then predator attention in these areas of high prey density may increase. For example, the highest skylark nest densities on arable land are often found on set-aside, but the predation rates of these nests are relatively high.
- If habitat change forces a species to nest or feed in vulnerable situations, predation may increase. For example, successful blackbird nests in hedgerows are less conspicuous than predated nests. If hedgerow management results in less dense nesting cover then this may result in increased predation rates for birds nesting in the hedge.
- If food shortage for chicks occurs then the chick begging intensity may increase and detectability of nests by predators may also increase. For example, cirl bunting chick survival was linked to chick condition (chicks with plenty of food survived better), although nests that failed were ultimately predated, suggesting that starving chicks were detected by predators more than healthier chicks. However, in this case, predators take chicks that may not have survived anyway.
- If mobile chicks are short of food then they need to move further to forage and may be more visible to predators as a result. For example, grey partridge chicks in areas of high pesticide use had to move greater distances which may have increased the risk of predation.

In these situations, land management that provides plenty of food and also improves the ability of prey to conceal themselves and their nests from predation is the most cost-effective solution (Plate 1.13). The judicious use of habitats to mitigate predation remains a potentially fruitful area for further research. Locally, predation by some species can also be reduced directly by legal methods of predator control, targeted at the period when game and songbirds are breeding.

Andy Hay/rspb-images.com

Plate 1.13 Ideally, lapwings rear their chicks in the field where they nest. If they have to move their chicks to areas with plenty of food, this may result in losses due to predation.

Table 1.5 A summary of the main predators of farmland birds. Here, main predators refer to species that are widespread in the UK and for which birds or their eggs are common components of their diet.

Main predators of farmland birds	Main prey items
Foxes	Eggs, flightless chicks, ground-nesting birds
Stoats, weasels, mink, badgers*, rats, hedgehogs*, crows and magpies	Eggs, flightless chicks
Sparrowhawks*	Adult birds and fledged juveniles

* Badgers, hedgehogs and sparrowhawks are protected under the Wildlife and Countryside Act 1981 and The Wildlife (Northern Ireland) Order 1985

Table 1.5 summarises the main predators of farmland birds in the UK. Under the terms of the Wildlife and Countryside Act 1981 (as amended by the Countryside and Rights of Way Act 2000) and The Wildlife (Northern Ireland) Order 1985 all wild birds, their nests, eggs and young are legally fully protected. However, Section 16 of the Wildlife and Countryside Act 1981 permits legal control of a few so called avian pest species through the use of General Licences. These species include the magpie, carrion crow, rook, jackdaw and jay. The General Licences permit specific methods to take and control various species on the above list under certain criteria and conditions of use. The General Licences are similar in all countries of the UK. Details can be found at http://www.defra.gov.uk/wildlife-countryside/vertebrates/gen-licence.htm.

At Loddington in Leicestershire, the Game Conservancy Trust improved the farmland habitat and undertook legal predator control, resulting in an increase in nesting success, a reduced rate of predation and an increase in breeding densities of nationally declining songbirds. The cause of the population increases could be due to the habitat improvement or the predator control or, more likely, a combination of the two.

Numbers of magpies and sparrowhawks have increased through the period when numbers of some farmland bird species have declined, but analysis of data from the Common Bird Census found no relationship between annual population changes of songbird species and the presence or absence of sparrowhawks or magpies.

Further reading

Evans K L (2004) The potential for interactions between predation and habitat change to cause population declines of farmland birds. Review paper. *Ibis* 146: 1–13.

Stoate C and Leake A (2002) *Where the birds sing. The Allerton Project: 10 years of conservation on farmland.* The Game Conservancy Trust, Fordingbridge.

Stoate C and Szczur J (2001) Could game management have a role in the conservation of farmland passerines: a case study from a Leicestershire farm. *Bird Study* 48: 279–292.

Tapper S C, Potts G R and Brockless M H (1996) The effects of an experimental reduction in predation pressure on the breeding success and population density of grey partridges *Perdix perdix*. *Journal of Applied Ecology* 33: 965–978.

Thompson D L, Green R E, Gregory R D and Ballie S R (1998) The widespread decline of songbirds in rural Britain does not correlate with the spread of their avian predators. *Proceedings of the Royal Society of London, Series B – Biological Sciences 265,* pp 2057–2062.

2 ECOLOGICAL REQUIREMENTS OF LOWLAND FARMLAND BIRDS

2.1 Grey partridge

2.1.1 Habitat use in summer

Grey partridges occupy arable crops, meadows, pastures and moorland fringes, but are predominantly found in open landscapes where there is an abundance of invertebrates for feeding chicks and seed food for adults, such as cereal field margins and open, weedy ground. They feed in arable crops, field margins, rough ground and unimproved grasslands. Breeding birds require thick, grassy vegetation for nesting and open ground for dusting and keeping chicks dry in wet weather. They nest in thick cover at the base of a short hedge or wall, on a low bank (such as a beetle bank) or in the crops themselves (but only where more suitable sites are unavailable). They tend to avoid banks or hedges with more than 10 trees per kilometre of linear cover. They prefer nest sites that are on a raised bank. Nest sites tend to be on free-draining soil on a slope, preferably facing south and with shelter from the prevailing wet weather.

Roger Wilmshurst/rspb-images.com

Plate 2.1
Grey partridge

2.1.2 Habitat use in winter

Grey partridges form coveys from July to February, which may consist of family groups or unsuccessful breeding birds. These birds are highly sedentary. Most will not move more than 3 km from their natal site. Their wintering requirements include seed-rich areas in which to forage, and cover to protect them from predators and weather. Although grey partridges can feed on other plant matter, they show a marked preference for seed-rich habitats such as winter stubbles, unimproved pastures, set-aside and wild bird cover. In these habitats, areas of tall stubble, grassy tussocks or wild bird cover also provide the cover they need.

2.1.3 Diet

Adults predominately eat plant matter including the leaves of cereals, grasses and clover, cereal grains and weed seeds. There is a preference for the larger weed seeds, especially those of the genera *Polygonum* (eg knotgrass, redshank) and *Fallopia* (black bindweed), but otherwise the seeds in the diet reflect their availability in the habitat. The proportion of weed seeds in the autumn diet fell from 31% dry weight in the 1930s to 4% in the 1970s. This was presumably due to the use of herbicides and changes in stubble management. The proportion of cereal grains in the diet rose during this period and the average bird weight did not alter. Insects form 5–12% of the adult diet in the summer.

In the first week after hatching, 95% of items eaten by chicks are insects, and the proportion of insects in the diet remains high during the second week. Chicks that are reared artificially grow and feather more quickly when insects are added to a diet of plant matter. Growth rate is an important survival factor for wild birds. In the wild, chicks show a clear preference for sawfly larvae which feed on grasses and cereal leaves, moth larvae which feed on grasses and weeds, larger bugs, weevils, leaf-beetles, small ground beetles and cereal aphids. They do not take insects that are very small (eg thrips), large (eg larger carabid beetles), swift-flying or below the soil surface.

2.1.4 Breeding

Breeding territories form when coveys break up in January or February. The territory size of both breeding pairs and wintering coveys is larger when food availability is lower.

The nest is a shallow scrape concealed in dense vegetation such as rank, tussocky grass, perennial herbs or arable crops. Old grass stems and dead leaves from the previous year's growth are used to construct the nest.

Egg-laying begins in late April/early May. They rear one brood per year but replacement clutches can prolong the breeding season to the end of August. The clutch consists of 10–20 eggs laid at one to two-day intervals. Incubation takes 23–25 days; the chicks leave the nest within a few hours after hatching and are able to fly 15 days after hatching. Chicks feed themselves from day one. The whole process from start of egg-laying to chicks fledging is therefore about eight weeks.

Predation is the main cause of nest loss. The eggs are taken mainly by foxes, carrion crows and magpies, but also by rats, hedgehogs and mustelids. The female will usually lay a replacement clutch if the nest is lost. However, if a fox, cat, stoat or rat kills the hen, this will obviously result in total loss of production. Chick survival is related primarily to chick food availability and weather.

2.1.5 Population trend and distribution

Bag records going back to the 1700s show the population trends of the grey partridge in Britain. There was a marked increase of at least four-fold between 1750 and 1880, probably due to a combination of increased nesting cover during the enclosures, extra food from legumes in the rotations and control of nest predators by gamekeepers. There was a decline in some areas, including Ireland, from 1914 probably due to reductions in predator control and cereal production. From the 1930s, and particularly since the 1960s, there has been a general decline due to agricultural intensification. This decline has been mirrored throughout the range in western Europe. The population declined by 86% in the UK between 1970 and 2001, but is since showing some signs of stabilising. The current population shows relatively high densities on the lighter soils of southern England and in the eastern UK, particularly from Norfolk to Grampian (Figure 2.1). Range contraction since 1970 has been largely from Ireland and western Britain.

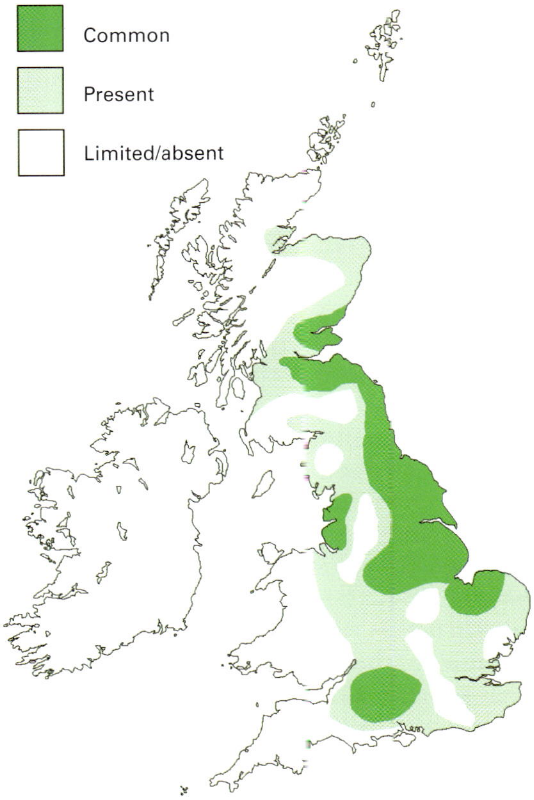

Figure 2.1
Distribution map showing the relative abundance of the grey partridge in Britain and Ireland in 1988–91.

2.1.6 Causes of decline

The Game Conservancy Trust has undertaken most of the research on this species. Chick survival was found to be the main factor regulating population size. Despite wide annual variation, average chick survival was generally stable before the 1950s and declined subsequently. Chick mortality is highest during the first two weeks after hatching. Chick survival is largely determined by the abundance of the preferred insects. In the Game Conservancy Trust study area in Sussex, chick survival rates were positively correlated with the abundance of insects within cereal crops in the summer. The survival of individual broods was positively correlated with the proportion of caterpillars, sawfly larvae, leaf beetles and weevils in the diet. Chick survival was estimated at over 40% before 1952 and around 30% after 1962, a change that was almost certainly due to the indirect impact of herbicides on the populations of these insects through destruction of their food plants. A further reduction to 22% was associated with the widespread use of broad-spectrum insecticides on cereals in summer. One favoured species in the 1930s was a weevil larva (*Phytonomous variabilis*), which is no longer available because its food plant, sainfoin, is now rarely grown. Subsequent research has shown that grey partridge broods occupy areas with relatively high densities of chick-food invertebrates, especially cereal field margins, and that brood sizes could be increased by not using summer insecticides or broad-spectrum herbicides on cereal headlands (conservation headlands).

The effects of chick survival rates (largely determined by availability of chick food invertebrates) and brood production rates (largely determined by nest predation) on the long-term population trend have been modelled on computer. The model shows that with low predation pressure the lower average chick survival rate that has occurred since the widespread use of herbicides on arable crops can maintain a stable population level, given the compensatory effects of the density-dependent variables of overwinter survival and brood production rates. However, if predator control is relaxed (ie predation pressure increases) then the effects of low chick survival and brood production rates lead to a population decline. The main predators are foxes, mustelids, hedgehogs, rats and crows (the carrion crow and the magpie).

A loss of nesting cover has exacerbated the decline. As field boundaries were removed to increase field sizes, many field margins were cropped up to the base of the hedge to maximise the cropped area, and hedge bottoms were sprayed out to prevent weed infestation of the crops.

2.1.7 Management advice

Arable

Computer modelling indicates that the optimal habitat requirements for grey partridges are a combination of 5% of arable area as insect-rich brood-rearing habitat and over 6 km/km^2 of linear nesting cover. Seed-rich habitats are also required for overwinter survival. Some ways in which such habitats can be created or improved are described below.

Create rough grassland in field corners and field margins of arable fields (see section 3.9) to increase food and nesting habitat for grey partridges. Select margins with short, thick hedges, stone walls, fence lines and ditches to provide the best nesting cover for grey partridges. Use a high proportion of cocksfoot in the seed mix to create a tussocky sward. Once established, the margins should be mown no more than once every three years to allow the tussocks to develop. Mowing should be done on a rotational basis so that some margins are always left uncut in any given area. If margins of 6 m or more are established then mow the 3 m abutting the crop annually in September to incorporate a more diverse grass structure. Partridges may nest in the tussocky vegetation but forage more in the shorter sward. The hedgerows (see section 3.1) themselves should be kept trimmed (preferably after the berry crop is over) to under 2 m in height so that they are not used as look-out posts by avian predators. Grey partridges like an open landscape, so avoid planting trees in areas that they occupy.

Beetle banks (see section 3.10) are 2 m grass strips created across fields. Their main purpose is to introduce overwintering habitat for invertebrates into the middle of larger arable fields (greater than 20 ha), but they may also be used by partridges as nesting cover and as a source of chick-food invertebrates in summer. A raised bank is established by ploughing in from both directions and then the grass strip is established and kept tussocky, ie not cut at all unless necessary to control weeds.

Conservation headland management (see section 3.12) (selective spraying) should be adopted on the outermost part ('headland') of cereal crops to encourage a sprinkling of broad-leaved weeds and an abundance of invertebrates that feed on them. To be successful, their management must follow Game Conservancy Trust guidelines. They can be from 6 m to 24 m in width, and can be undertaken in conjunction with grass margins and beetle banks. In some situations, low-input cereal crop management can be adopted to create the benefits of conservation headlands over the whole field.

Alternatively, insect-rich foraging areas can be provided by creating cultivated, uncropped field margins to allow arable plants and their associated insects to flourish, or by establishing pollen and nectar mixtures. The practice of cutting pollen and nectar mixtures in June to promote late flowering for insects may jeopardise nesting attempts, so avoid siting these in areas where grey partridges are likely to nest.

On the rest of the cereal-growing area, use insecticides, molluscicides, fungicides and herbicides only where the problems exceed the economic threshold, and then use a selective chemical if one is available (eg for aphids, grass weeds and cleavers). Insecticides used after 15 March pose a particular threat to grey partridges as they remove invertebrates such as predatory insects and spiders that have overwintered in the grass margins and moved out into the crop in the spring.

Spray and cultivate stubbles (see section 3.14) as late as possible. If the stubble can be left over winter then it will provide an important winter feeding habitat. Undersown stubbles that are left to develop into a grass ley provide a very rich source of sawfly larvae for partridge chicks in the following summer (see section 3.18). The management of the preceding cereal crop largely determines the benefit of winter stubbles as feeding habitat. The seed food it provides will be cereal grains dropped during harvesting and seeds of weeds that germinated with the preceding crop. If the weed flora was overzealously controlled then the seed food available will be limited. Spring cereal stubbles often hold more seed food as the herbicide regime is less intensive and many of the important seed food sources for birds are from spring-germinating plants such as the polygonums. Conservation headland management in the preceding crop will boost the seed food available in the stubble margin.

Wild bird cover (also known as wild bird seed mixture or unharvested crops) (see section 3.7) can provide an abundant source of seed and cover through the winter. Establish a mix of a cereal and a brassica in April/May and leave until the following March. A two-year mix incorporating kale can also be used as nesting cover, as can late-autumn sown millet and oilseed rape. The artificial provision of food (wheat grain) using feeders is also a possibility, in which case they should be placed in areas already frequented by coveys, near cover (which can be created using brush from hedgerow or woodland management).

On rotational set-aside (see section 3.15), natural regeneration provides a rich seed source until the point when management by spraying, cutting or cultivation is necessary. This may be used as nesting cover if the vegetation is thick enough. Ideally, spraying should be delayed until at least mid-July and cultivation delayed until mid-August. If early spraying is required to control a significant weed problem, use of a selective chemical or spot spraying may prolong the benefits throughout the breeding season. If a non-selective herbicide is used, leave the cover to die off over a period of weeks; do not cut or cultivate the cover during the breeding season.

Non-rotational set-aside strips may be used for nesting if a tussocky grass cover is established (eg using a mixture containing cocksfoot, which will take two years to become tussocky) and left unmanaged until at least August. Cut as close as possible to the 15 August deadline and make use of the option to leave 25% of the non-rotational set-aside area uncut for up to three years. This can be rotated around the field.

An arable estate in France with very high grey partridge densities uses the following recipe with manifest success:

- patchwork quilt of different crop types
- conservation headlands with no fertiliser
- wild bird cover strips on 20 m set-aside, often across the middle of fields
- no summer insecticides on the outer 12 m of crop
- year-round feeders containing wheat (two per pair of grey partridges)
- few trees
- predator control

Grassland

In purely pastoral areas, grey partridges would benefit from the introduction of any form of arable cropping, such as whole-crop cereals to boost the availability of seeds (see section 3.18). Grey partridges are often found on unimproved and semi-improved pastures that provide a greater diversity and abundance of invertebrates. Such grasslands should be managed to maintain this interest. Even on improved grass, allowing the margin to grow and seed without grazing or cutting may create nesting and feeding habitat (see section 3.24). The diversity of plants and invertebrates will be improved by not fertilising or reseeding these margins. Cutting or grazing between September and February will prevent the encroachment of the hedge. This may be undertaken annually, or once every two or three years to allow a more tussocky margin to develop.

Can release of hand-reared birds help?

In general, the release of hand-reared birds as carried out for shooting purposes will not help the conservation of grey partridges. Typically, hand-reared birds do not behave naturally in the wild: they are very vulnerable to predation and rarely breed successfully. Moreover, the primary cause of decline is the loss of suitable habitat on farmland, so restoration of suitable habitat must be the priority. In cases where partridges have failed to respond to habitat restoration, possibly because their numbers are so low that they are caught in a predation trap or they are extinct in the area, artificial re-establishment may be necessary. The Game Conservancy Trust is currently researching the optimal methods for re-establishing grey partridges through releasing, with the aim of producing science-based guidelines by the end of 2006. Until the research has concluded, releases are not recommended. It would be inadvisable to introduce the grey partridge to areas outside the historical range for this species as, from a conservation perspective, they would be highly unlikely to be successful.

Table 2.1 summarises the beneficial options.

Table 2.1 Effects of different forms of land management on the grey partridge.

	Nesting	Insect food	Seed food
Arable pockets in pastoral areas	✔	?	✔
Unimproved grassland		?	?
Hedges	?		
Ditches	?	?	
Arable field margins and corners	✔	✔	
Non-rotational set-aside	?	✔	
Cultivated margins		?	?
Improved grass field margins	?		
Beetle banks	✔	✔	
Conservation headlands		✔	?
Low-input spring cereals		✔	
Broad-spectrum pesticides		✘	✘
Rotational set-aside	?	?	✔
Overwintered stubble			✔*
Undersown spring cereal		✔	
Pollen and nectar mixtures	✘?	?	
Wild bird seed mixture	?	?	✔
Tree planting	✘		

✔ indicates positive benefits
? indicates potential benefits/impacts dependent on management and outcome
✘ indicates negative impacts
* especially if preceded by low-input cereal

Further reading

Aebischer N J (1997) Gamebirds: management of the grey partridge in Britain. In M Bolton (ed) *Conservation and the Use of Wildlife Resources*, pp 131–151. Chapman and Hall, London.

Aebischer N J and Ewald J A (2004) Managing the UK grey partridge *Perdix perdix* recovery: population change, reproduction, habitat and shooting. *Ibis* 146 (Suppl. 2): 181–191.

Aebischer N J and Potts G R (1998) Spatial changes in grey partridge (*Perdix perdix*) distribution in relation to 25 years of changing agriculture in Sussex, UK. *Gibier Faune Sauvage* 15: 293–308.

Bro E, Reitz F, Clobert J, Migot P and Massot M (2001) Diagnosing the environmental causes of the decline in grey partridge *Perdix perdix* survival in France. *Ibis* 143: 120–132.

Chiverton P A (1999) The benefits of unsprayed cereal crop margins to grey partridges *Perdix perdix* and pheasants *Phasianus colchicus* in Sweden. *Wildlife Biology* 5: 83–92.

GCT (2001) *Conserving the Grey Partridge*. The Game Conservancy Trust, Fordingbridge. [GCT partridge leaflet approved by BAP group]

GCT (2004) *Factsheet 2. Guidelines for the Management of Field Margins: recommendations for the management of field margins, incorporating the field boundary, boundary strip and Conservation Headlands*. The Game Conservancy Trust, Fordingbridge.

Green R E (1984) The feeding ecology and survival of partridge chicks (*Alectoris rufa* and *Perdix perdix*) on arable farmland in East Anglia. *Journal of Applied Ecology* 21: 817–830.

Potts G R (1986) *The Partridge. Pesticides, Predation and Conservation*. Collins, London.

Potts G.R. 1997. Cereal farming, pesticides and grey partridges. In D J Pain and M W Pienkowski (eds) *Farming and Birds in Europe: A Common Agricultural Policy and its Implications for Bird Conservation*. Academic Press, London.

Potts G R and Aebischer N J (1991) Modelling the population dynamics of the grey partridge: conservation and management. Pp 373–390 in C M Perrins, J-D Lebreton and G J M Hirons (eds) *Bird Population Studies: Their Relevance to Conservation Management*. Oxford University Press, Oxford.

Potts G R and Aebischer N J (1995) Population dynamics of the Grey Partridge *Perdix perdix* 1793–1993: monitoring, modelling and management. *Ibis* 137: S29–S37.

Rands M R W (1985) Pesticide use on cereals and the survival of grey partridge chicks: a field experiment. *Journal of Applied Ecology* 22: 49–54.

Rands M R W (1986) Effect of hedgerow characteristics on partridge breeding densities. *Journal of Applied Ecology* 23: 470–487.

Rands M R W (1986) The survival of gamebird (Galliformes) chicks in relation to pesticide use on cereals. *Ibis* 128: 57–64.

Sotherton N W S (1991) Conservation Headlands: a practical combination of intensive cereal farming and conservation. Pp 373–397 in L G Firbank, N Carter, J F Darbyshire and G R Potts (eds) *Ecology of Temperate Cereal Fields*. Blackwell Scientific Publications, Oxford.

Tapper S C, Potts G R and Brockless M H (1996) The effect of an experimental reduction in predation pressure on the breeding success and population density of grey partridges *Perdix perdix*. *Journal of Applied Ecology* 33: 965–978.

2.2 Lapwing

2.2.1 Habitat use in summer

Breeding lapwings need all-round visibility and good quality foraging areas. The main factors in site selection are vegetation height and density, and accessibility of surface and sub-surface invertebrates. Lapwings prefer short swards for nesting. They will tolerate swards up to 15 cm high in sparse arable crops or unimproved grassland, but may require shorter swards in thicker vegetation, such as improved grasslands. Short, unimproved pasture with scattered tussocks to provide some cover is the ideal grassland habitat. Rough, cultivated fallow is the best arable nesting habitat. Lapwings generally avoid nesting in enclosed fields of less than 5 ha or close to field boundaries that can harbour predators, or in an area in the immediate vicinity of trees or other features that can act as predator perches.

The 1987 BTO lapwing survey estimated that 95% of lapwings in England and Wales bred on farmland, with 56% breeding on grassland and 39% on arable farmland. Lapwings breed in grazed pastures, spring-tilled arable land and occasionally in hay fields (depending on the structure). Selection in relation to habitat availability shows there is a clear preference for spring tillage and rough grazing. Improved grass leys and autumn tillage are generally avoided. Lapwings are generally found in higher densities on mixed farmland than on purely arable or pastoral farmland. However, very high densities are found on unimproved or semi-improved grassland systems at sites such as the Pennine Dales, Islay and the North Kent Marshes. They also breed in semi-natural habitats: wetlands, heathland, moorland, chalk and upland grasslands. In all situations, they show a strong association with grazing livestock. Where available, they select moist soils where there is easy access to soil invertebrates through the summer, although they also nest on mixed farmland on chalk or on Breckland, far away from any water. Birds nesting on spring-tilled arable land generally require open, grazed pasture on adjacent or nearby fields to rear chicks. This may be because the arable crop will grow too tall to maintain all-round visibility, or because of greater availability of invertebrates in the grassland. Birds breeding on fallow plots will also move to grazed pasture if it is available, although this seems to be less important than for pairs nesting in a spring crop because they are not driven out by crop growth.

Andy Hay/rspb-images.com

Plate 2.2 Lapwing

2.2.2 Habitat use in winter

In winter, lapwings are attracted to grazed pasture, winter stubbles and newly ploughed fields. The British Isles is an important wintering area for lapwings breeding in Scandinavia. Birds breeding in the UK generally move shorter distances, but severe winters cause hard-weather movements southwards and westwards into Ireland. Many people regard the large winter flocks as an indication of good local breeding populations, as they do not realise the extent to which this species migrates.

Lapwings tend to feed and roost in large fields to improve detection of predators. They show a clear preference for feeding in grassland over arable land in the winter, especially in late winter. They prefer to feed in fields where the vegetation is less than 75 mm high. Ploughed fields are preferred for roosting.

2.2.3 Diet

Lapwings feed on bare ground or vegetation less than 15 cm high with access to the soil. They feed in areas with high densities of surface or sub-surface invertebrates, especially beetles and their larvae, earthworms, and other large soil invertebrates such as leatherjackets. They also take flies, grasshoppers, moth larvae, ants, and larvae of other Hymenoptera, caddis flies, mayflies, dragonflies, spiders, molluscs, millipedes, harvestmen and woodlice. They take a small amount of plant matter including grass leaves and seeds.

The importance of earthworms in the diet relates to availability governed by soil conditions and climate. Supply may be limited in hot, dry summers, especially on arable and dry grassland. Dung fauna may be an important food source on grazed pasture in some situations, and so the impact of worming treatment on dung fauna may be an issue. Soil invertebrates tend to be more important in the early part of the breeding season, whilst surface invertebrates increase in importance as the ground becomes harder. Earthworms are the most important component of the winter diet. Large winter feeding flocks are an indication of large earthworm densities.

2.2.4 Breeding

Most lapwings return to within 5 km of their natal area to breed. Birds start arriving at breeding grounds from mid-January. Most birds breed in the first season after fledging. Males can be polygamous.

Pairs either breed singly or in loose colonies of up to 30 pairs. Birds nesting in colonies can reach densities of up to six pairs per hectare. There is some evidence that breeding success by pairs that nest in loose colonies is higher than those which nest singly under some circumstances. Where better breeding success is observed, this is probably due to the more effective detection and mobbing of predators (particularly gulls and crows) than can be achieved by a single pair.

The nest is built in the open, often on raised ground or a ridge that gives maximum visibility. The nest is a scrape created by the adults and variably lined with vegetation.

Eggs are generally laid from late March to early June. Chick rearing generally coincides with peak food availability. The egg-laying period may be truncated on arable land by crop growth inhibiting the visibility of a sitting bird. Nesting attempts started in the latter period of the breeding season tend to be less successful, probably due to food shortages as earthworms burrow deeper and insect larvae hatch.

The clutch is usually of four eggs, rarely two to five. Small clutches may result from egg loss due to agricultural operations during egg-laying, or predation. Eggs are laid at a rate of one per day, incubation is 25–28 days and chicks fledge 35–40 days after hatching. The complete breeding cycle for a brood of four therefore takes about 10 weeks from the start of egg-laying.

Lapwings only rear one brood per year but may lay one or two (exceptionally three or four) replacement clutches if an attempt fails. The period between nest loss and starting a replacement clutch is about eight days.

Arable nesting pairs are prone to nest destruction by cultivation and rolling operations. Breeding pairs may be forced to abandon nests or walk chicks out of a crop as it grows too tall or dense. Fledging success is generally lower in cases where chicks are required to move further. In grassland, nests may be destroyed during harrowing, rolling or muck spreading operations, or by being trampled by stock. Cultivation or harrowing during egg-laying may lead to females completing clutches in new nests and therefore reduce the average clutch size on the field. In all locations, predation, especially by crows, gulls or foxes, or severe weather may be significant causes of breeding failure. Hot or dry summers can reduce food availability, and cold or wet weather can kill young chicks.

An analysis of ringing recoveries estimated annual adult survival rates of 75%, and first year survival rates of 60%, but showed a strong dependence on winter weather conditions. Analysis of adult survival indicates that each pair needs to rear 0.6 chicks per year to maintain the population.

2.2.5 Population trend and distribution

Historically, lapwings have experienced pronounced population fluctuations, some of which result from harsh winters, and others probably reflect changes in land-use that affect breeding success. These fluctuations also show that the lapwing has a high capacity to recover quickly and recolonise areas after a decline. In Britain, there was a marked decline through the end of the 19th and the early 20th centuries when egg-collecting as well as land-use changes probably had an impact. There was a recovery following the Lapwing Act of 1926, which offered some protection, but a widespread decline again from the 1940s. Numbers were stable through the 1960s and 70s. There was a 41% decline in the UK between 1970 and 2001 which was most pronounced in the 1980s. This decline may have been initiated by a series of harsh winters that reduced overwinter survival rates. However, recent studies have indicated that survival rates have been high and areas showing marked declines have breeding productivity below the critical 0.6 chicks reared per pair (the level of productivity required to maintain a population) on either grassland or arable systems. Nest records indicate that there has been a decrease in nest survival.

The BTO lapwing surveys of 1987 and 1998 provided a more thorough census of the English and Welsh populations and showed a 49% decline in this 11-year period, with declines greatest in Wales and south-west England. Surveys in Northern Ireland estimated a population decline of 66% between 1987 and 1999. The distribution map shows that the species was still widespread in the early 1990s, but that concentrations were generally low in the Highlands, Ireland and south of a line from the Dee to the Wash in England and Wales (Figure 2.2). Between 1970 and 1990, range contractions occurred mainly in Ireland, Wales and south-west England. More recent data from the Breeding Bird Survey (since 1994) indicate that the greatest losses are now occurring in Scotland.

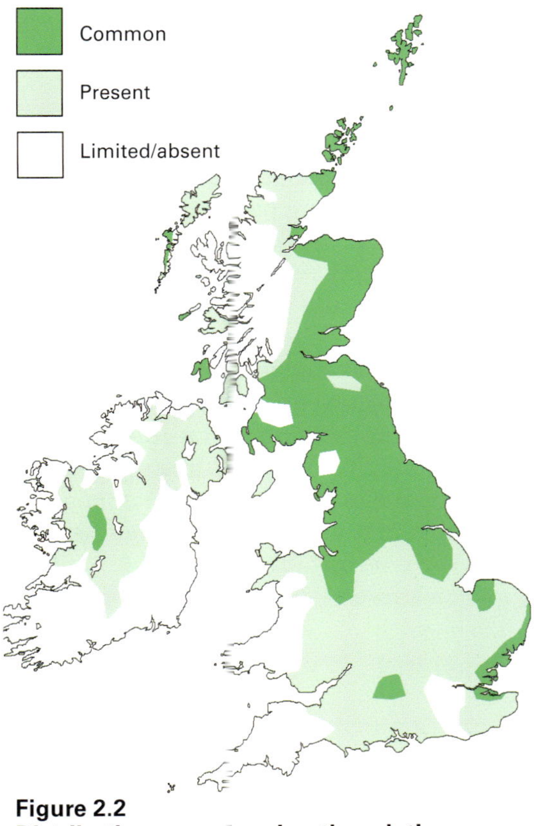

Figure 2.2
Distribution map showing the relative abundance of the lapwing in Britain and Ireland in 1988–91.

2.2.6 Causes of decline

The annual survival rates of adult and first year birds are generally at least as high during periods of decline as they are at other times, indicating that breeding season factors may be responsible for the decline. Some land management changes have had dramatic effects on the availability of nesting habitat for lapwings. Land management changes that have caused loss of nesting habitat include:

- shift from spring to autumn sowing of arable crops reducing nesting opportunities on arable land
- loss of wet grassland through field drainage
- faster grass growth through the use of inorganic fertiliser and use of high yielding varieties of grasses
- loss of invertebrate-rich grassland through reseeding

- loss of mixed farming which allowed birds to nest in spring crops and move chicks to grazed pasture: this has had a pronounced effect on habitat suitability for arable nesting birds.

Nest failure rates have increased since the mid-1970s and may have contributed to recent declines. These are at least partially related to changes in land management. For example, greater grass productivity has led to the switch from hay to silage, increasing the nest losses during mowing, and higher stocking rates on grazed pasture, which increases the risk of nest trampling. Other major factors reducing breeding success are predation, farming operations and crop growth that, in addition to the above factors, can exacerbate rates of decline. BTO nest record card data from 1962 to 1985 showed a 47% nest loss on grassland (10% desertion, 20% predation, 9% farm operations and 8% stock trampling) and 39% on arable (4% desertion, 9% predation and 26% farm operations).

2.2.6 Management advice

Arable

Spring drilled crops (drilled in February or later), especially those adjacent to grazed pasture, provide the right nesting habitat for lapwings. Where spring crops are not a viable option then a lapwing plot created on set-aside is a suitable alternative on arable land (see below). Nests on land going into spring crops are at risk of cultivation, drilling and rolling operations after mid-March. Spring cultivations and drilling should be carried out as early in the season as possible, so that early nesting birds are able to re-nest, although 'spring crops' drilled before February are usually too advanced by the time the nesting season starts. Late rolling of crops should be avoided to prevent the loss of nests close to hatching, and result in failures when the crop is too advanced to allow re-nesting.

Look out for lapwing nests during cultivation, drilling or rolling operations after mid-March and work around them where possible. If a series of operations are required to establish a crop after mid-March then attempt to complete them within a week so that only one breeding attempt per pair is disrupted. Spraying operations are unlikely to cause a problem. Maize and other late-sown crops can provide useful nesting habitat for pairs that have failed nearby. However, these crops are particularly likely to attract breeding lapwings during establishment operations, which may result in total breeding failure. The above guidelines are particularly important in these crops.

Any measures to boost invertebrate food within arable fields used by nesting lapwings, such as field margin (see section 3.9) management and beetle banks (see section 3.10), should improve foraging opportunities for them. In large arable fields, beetle banks may also provide cover to protect chicks from predators. Wherever there is an opportunity to create grazed grassland in purely arable landscapes, this will greatly enhance the quality of the habitat for lapwings. Spring crops should be rotated around adjacent fields, or a cultivated fallow plot should be sited on an adjacent field.

Create summer fallow plots (see section 3.16) using agri-environment schemes or on set-aside (with a derogation from the relevant government department). In fields with a history of breeding lapwings, seek to cultivate the field or a 2 ha plot in the centre. This should be ploughed in the autumn and allowed to weather down overwinter, or ploughed and disced in February. The field or plot should be left unmanaged until July. Where use by lapwings is unknown, choose a large open arable field adjacent to a grazed pasture. Plots can be created on a rotational or non-rotational basis, although a fixed plot in a suitable location is better as the birds are more likely to return year after year, and there is no risk of a weed burden or wheat bulb fly problem affecting the following crop.

Grassland

Any remaining unimproved grassland, especially undrained, wet grassland, should be maintained (see section 3.19). This may be funded by agri-environment schemes. High water tables and small areas of surface water, such as open, shallow-sided drainage channels, are beneficial. Where possible, they should be maintained until late June. The creation of wader scrapes (see section 3.22) to maintain areas of damp ground through the summer is an option in some agri-environment schemes. However, extensive areas of surface water from early May onwards will encourage the spread of soft rush and therefore be detrimental to the long-term management of the site. Topping after birds have finished breeding can be used to manage rush encroachment (see section 3.23). Any small-scale areas of bare soil created by moles, poaching by cattle or surface water are also beneficial. Chain harrowing and rolling, where necessary, should not be carried out between mid-March and the end of June as this may destroy nests.

The banks of ditches (see section 3.4) and foot-drains can provide ideal foraging habitat for lapwings and their chicks through the summer if they are open and accessible. Shallow slopes with open access for livestock grazing can provide ideal foraging habitat. The shallowest slope, and the shortest sward achievable, provide the best habitat. Lapwing chicks searching for invertebrates often use poached ground at the edge of water, although excessive poaching leads to water pollution problems.

Grazing from June onwards should be managed to provide a short sward (5 cm) for nesting lapwings in the following March. Rushes and other tall vegetation should be limited in height (less than 10 cm) and extent by cutting or grazing before mid-March to avoid obscuring the visibility of nesting birds. Small-scale rush infestation is associated with high breeding lapwing densities. This may provide cover or merely be correlated with the presence of wet grassland with high invertebrate availability. However, breeding lapwing densities fall off when rush cover exceeds 30%.

Livestock grazing presents a risk of nest trampling, but may be essential if the sward would otherwise exceed an average of 8 cm and become too dense. If a sward can remain suitable by shutting up for a minimum of six weeks from late March then this is ideal. Hay, and even silage management, can provide suitable nesting habitat in some locations, provided the chicks can fledge or move out of the field prior to cutting. However, in most improved grasslands, especially in lowland situations, grass would become too tall and thick. This would not enable effective feeding and predator evasion, and the adults would be unable to keep young chicks warm and dry in wet weather. If grazing has to be carried out during the breeding season then reduce the stocking levels to that required to maintain a sward of less than 15 cm.

Introducing spring-sown arable fodder crops into pastoral areas (see section 3.18) will provide potential nesting habitat. Look out for nests during operations after mid-March.

General

Disturbance during bad weather may lead to the chilling of eggs or young chicks. Do not plant new hedgerows, woodlands or hedgerow trees in areas used by breeding lapwings, and maintain existing hedgerows without allowing a significant increase in height.

Table 2.2 summarises the beneficial options.

Table 2.2 Effects of different forms of land management on the lapwing.

	Nesting	Invertebrate food
Spring-sown crops	✔	
Fallow nesting plot	✔	✔
Maintain grazing on unimproved grass	✔	✔
Maintain pasture with high water levels	✔	✔
Rush management	✔	✔
Arable reversion to grassland	?	✔
Hay meadows	?	?
Creation of wader scrapes		✔
Grazed pasture adjacent to arable fields		✔
Ditches (with shallow banks)		✔
Arable field margins and corners		?
Beetle banks		?
Hedge and tree planting	✘	

✔ indicates positive benefits
? indicates potential benefits dependent on management and outcome
✘ indicates negative impacts

Further reading

Baines D (1990) The role of predation, food and agricultural practice in determining the breeding success of the lapwing *Vanellus vanellus* on upland grasslands. *Journal of Animal Ecology* 59: 915–929.

Berg A, Lindberg T and Kallebrink K G (1992) Hatching success of lapwings on farmland – differences between habitats and colonies of different sizes. *Journal of Animal Ecology* 61: 469–476.

Catchpole E A, Morgan B J T, Freeman S N and Peach W J (1999) Modelling the survival of British lapwings *V. vanellus* using ring-recovery data and weather covariates. *Bird Study* 46 (supplement): S5–13.

Fleming J (2001) *Farmland waders in Scotland*. The RSPB, Edinburgh.

Galbraith H (1988) Effects of agriculture on the breeding ecology of lapwings *V. vanellus*. *Journal of Applied Ecology* 25: 487–503

Green R E (1985) *The Management of Lowland Wet Grassland for Breeding Waders*. The RSPB, Sandy.

Peach W J, Thompson P S, and Coulson J C (1994) Annual and long-term variation in the survival rates of British lapwings *V. vanellus*. *Journal of Animal Ecology* 63: 60–70.

Pickup T (1999) *Lapwing reecovery project management manual*. The RSPB, Bangor, Gwynedd.

Sheldon R D (2002a) Breeding success and chick survival of Lapwings *Vanellus vanellus* in arable landscapes: with reference to the Arable Stewardship Pilot Scheme. Unpublished PhD thesis, Harper Adams University College and Open University.

Sheldon R D (2002b) Lapwings in Britain – a new approach to their conservation. *British Wildlife* 14: 109–116.

Shrubb M (1990) Effects of agricultural change on nesting lapwing *V. vanellus* in England and Wales. *Bird Study* 37: 115–127.

Shrubb M and Lack P C (1990) The numbers and distribution of lapwings *V. vanellus* in England and Wales in 1987. *Bird Study* 38: 20–37.

Taylor R (1999) *Lapwing training manual*. The RSPB, Huddersfield.

Thompson P S, Baines D, Coulson J C and Longrigg G (1994) Age at first breeding, philopatry and breeding site-fidelity in the lapwing *V. vanellus*. *Ibis* 136: 474–484.

Thompson P S and Coulson J C (1994) Population dynamics of the lapwing *V. vanellus* in stable and declining populations. Unpublished report to NERC.

Tucker G M, Davies S M and Fuller R J (eds) 1994. *The ecology and conservation of lapwings* Vanellus vanellus. *UK Nature Conservation No 9*. Joint Nature Conservation Committee, Peterborough.

Wilson A M, Vickery J A and Browne S J (2001) Numbers and distribution of Northern Lapwings *Vanellus vanellus* breeding in England and Wales in 1998. *Bird Study* 48: 2–17.

2.3 Turtle dove

2.3.1 General ecology

Turtle doves overwinter in sub-Saharan Africa and breed throughout most of Europe. After migrating through Spain and France they arrive in Britain from late April and start breeding immediately. After breeding, turtle doves start the autumn migration in early August, although birds can start nesting attempts throughout August. In Britain, the turtle dove favours the relatively warm, dry conditions of southern and eastern England. Turtle dove habitat consists of a mixture of tall, thick, mature hedgerows, shrubby woodland margins or tall scrub for nesting, and open, weedy patches for feeding.

2.3.2 Diet

The turtle dove's diet consists solely of the seeds of weeds and crops obtained by foraging on the ground. Turtle doves feed at a range of sites, including farmyards, stubbles after harvest, animal feed sites and weedy areas, especially where the vegetation is short and sparse. They forage over relatively large distances, and have been recorded feeding over 10 km from nesting sites. There has been a marked change in diet since the 1960s, when over 90% of the diet was made up of weed seeds. In the 1990s, weed seeds composed less than 40% of the diet, with wheat and oilseed rape seeds making up the rest. Given the relative abundance of weed and crop seed, turtle doves clearly selectively eat weed seeds where these are available. The change in diet reflects the availability of feeding habitats, which in the 1960s included more weedy arable fields, clover leys and hay fields; in the 1990s, these were either absent or greatly reduced. The reduced amount of weeds within cropped areas means that weed seeds are now largely confined to headlands and uncropped areas (including set-aside). The weed seeds commonly taken include fumitory (seed availability peaks in June), and chickweed and knotgrass (seed availability peaks in August). Seeds are generally more available during the second half of the breeding season and feeding sites change through the summer as seed sources become depleted and new sources become available. A single rich food source, such as set-aside, a particularly weedy crop or an area of spilt grain, can attract most of the turtle doves in an area if feed sites are very limited.

Ernie Janes/rspb-images.com

Plate 2.3 Turtle dove

Chicks are fed on crop milk produced by the adult for the first few days, and by about seven days they are fed on the same diet as the adults. There is no evidence that reduced seed availability has affected the success of individual nesting attempts or chick condition.

2.3.3 Breeding

Turtle doves breed in a range of habitats including young woodland, plantations and heathland, but the majority nest in arable areas with a mix of suitable nesting and feeding habitats. A study in East Anglia in the 1990s found 42% of nests in hedgerows, 25% in scrub, 18% in young plantations and woodland, and 15% in isolated bushes. The woody species used for nesting were hawthorn in 53% of cases, Norway spruce in 17%, elder in 14% and apple in 6%. Nest location ranged from just above ground level up to 20 m high, but averaged 2 m above ground. Many nests were closely associated with climbing plants such as traveller's joy, honeysuckle and bramble. The availability of nesting habitat could be the factor limiting breeding densities in some areas. Hedgerows used were generally over 4 m high and about 3 m wide.

Turtle doves breed in the first summer after fledging. Egg-laying starts in early-May and finishes by the end of August. Peak breeding activity is in the first two weeks of June. In the 1960s there was a second peak in early August, but this no longer occurs.

The clutch typically contains two eggs, ranging from one to three. Incubation generally takes 13 to 16 days and chicks fledge after about 20 days, so the full cycle takes five to six weeks. Turtle doves can rear three broods in a year if suitable conditions exist, although only about 20% of pairs started a new nesting attempt after successfully rearing a first brood in East Anglia in the 1990s. Multiple broods were more common 30 years ago. They make up to five nesting attempts if earlier attempts fail.

In the 1960s each pair laid on average just under three clutches of eggs, and produced about two fledged young. In the 1990s this had decreased dramatically to about one and a half clutches per pair and just over one fledged young.

2.3.4 Population trend and distribution

Historically the range of turtle doves in the British Isles has been restricted to England and Wales. There is evidence that turtle dove numbers in Britain increased between the mid-1800s and 1970. However, they declined by 77% between 1970 and 2001. The range contracted by 25% in the 1970s and 1980s, especially from Wales (where it is now very rare) and south-west England (Figure 2.3).

2.3.5 Causes of decline

A number of factors have probably contributed to the decline of the turtle dove. Today, birds finish nesting earlier and only produce half the number of clutches and half the number of fledged chicks per pair compared with the 1960s. This is sufficient to explain the decline that has occurred over this period. Ironically, seed food availability in the latter period of the breeding season is greater than at the start of the season, due to the availability of ripening rape seed and cereal grains and it may be that adults arriving in the UK in late April struggle to attain breeding condition, and rearing a first brood reduces their fitness to such a degree that they are incapable of attempting subsequent broods. The switch from feeding on weed seeds in weedy crops scattered across arable farmland in the 1960s to scarce seed-foraging opportunities such as sparse crops, grain spillages or weedy field corners in the 1990s has probably caused the reduction in the number of breeding attempts by reducing adult fitness.

Loss of mature hedgerows and scrub on farmland may have limited nesting habitat in some areas. It is unknown whether any additional factors have affected survival of

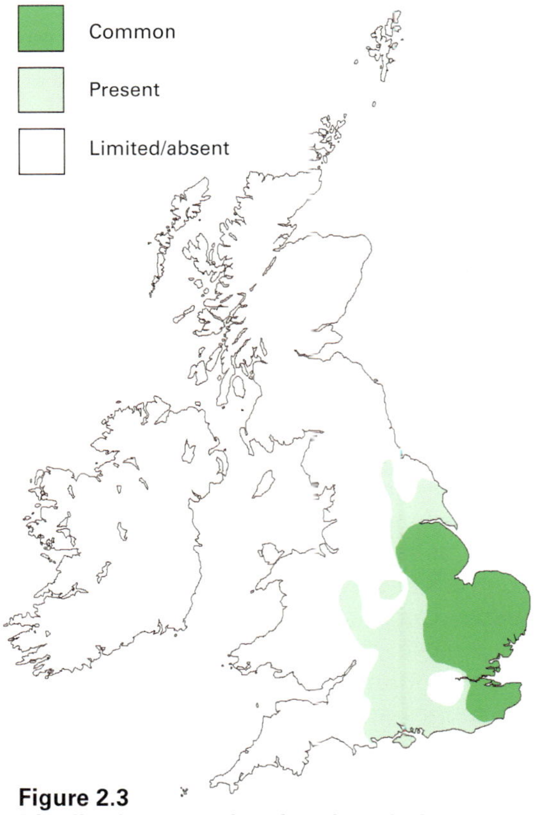

Figure 2.3
Distribution map showing the relative abundance of the turtle dove in Britain and Ireland in 1988–91.

birds in their African wintering grounds or on migration.

2.3.6 Management advice

Arable

The overall objective on arable farmland is to provide a rich seed-source in combination with good nesting habitat (see 'hedges, scrub and woodland' below). Easy access to the ground for feeding is important for this species, so dense crops are not useful, even if seeds are plentiful. Cultivated field margins (see section 3.11) and summer fallows (see section 3.16) can attract turtle doves if a diverse broad-leaved flora develops rather than a thick sward of competitive grass weeds and cleavers. Conservation headlands and low-input cereals (see sections 3.12 and 3.13) can provide areas of useful broad-leaved weeds, especially if no fertiliser is used, as the birds will have easier access to the ground. Cereal, pea and rape stubbles left until the end of August will provide important feeding sites for birds prior to migration.

Natural regeneration of rotational set-aside (see section 3.15) can provide weed seeds if the cover is not sprayed off too early in the spring, or does not become too tall and dense. Wild bird cover (see section 3.7) sown with a low seed rate is beneficial, especially if weed seeds such as fumitory are present. The low seed rate allows germination of weeds and gives the birds easier access to the ground. Oilseed rape and cereals are useful components in the mix.

Grassland

Restoration of species-rich hay meadows and herb-rich grasslands will provide alternative feeding habitats. Arable pockets (see section 3.18) or wild bird cover (see section 3.7) in pastoral areas are essential to maintain turtle doves in areas that have become predominately agricultural grasslands.

Hedges, scrub and woodland

Selected mature hedgerows (see section 3.1) should be allowed to reach at least 4 m in height and exceed 2 m in width. Cutting should be no more frequently than once every two to three years, preferably trimming alternate sides of the hedge in different years. Climbers, such as bramble, traveller's joy and honeysuckle, should not be removed from mature hedgerows.

Areas of scrub and scrubby margins of woodlands should be retained. New and existing woodland margins or field corners should be planted with hawthorns (but do not plant trees on species-rich grasslands). New plantings should be sited close to seed-rich foraging habitats wherever this does not detract from the requirements of farmland species requiring open arable farmland, such as grey partridge, lapwing and corn bunting.

Table 2.3 below shows beneficial options.

Table 2.3 Effects of different forms of land management on the turtle dove.

	Nesting	Seed food
Maintain tall, thick hedges	✔	
Scrub and scrubby wood margins	✔	
Arable pockets in pastoral areas		✔
Summer fallow		✔
Cultivated margins		✔
Hay meadow		✔
Rotational set-aside		?
Wild bird cover		?
Conservation headlands		?
Low-input cereal		?
Post-harvest stubble		?
Unimproved grassland		?

✔ indicates positive benefits
? indicates potential benefits dependent on management and outcome

Further reading

Browne S J and Aebischer N J (2001) *The role of agricultural intensification in the decline of the turtle dove* Streptopelia turtur. English Nature Research Report No 421, Peterborough.

Browne, S J and Aebischer, N J (2002) The effect of supplementary feeding on territory size, territory density and breeding success of the Turtle Dove *Streptopelia turtur:* a field experiment. *Aspects of Applied Biology* 67: 21–26.

Browne S J and Aebischer N J (2003) Habitat use, foraging ecology and diet of Turtle Doves *Streptopelia turtur* in Britain. *Ibis* 145: 572–582.

Browne S J and Aebischer N J (2004) Temporal changes in the breeding ecology of European Turtle Doves *Streptopelia turtur* in Britain, and implications for conservation. *Ibis* 146: 125–137.

Browne S J, Aebischer N J, Yfantis G and Marchant J H (2004) Habitat availability and use by Turtle Doves *Streptopelia turtur* between 1965 and 1995: an analysis of Common Birds Census data. *Bird study* 51: 1–11.

Calladine J R, Buner F and Aebischer N J (1997) *The Summer Ecology and Habitat Use of the Turtle Dove* Streptopelia turtur: *A Pilot Study*. English Nature Research Report No 219, Peterborough.

Murton R K (1968) Breeding, migration and survival of turtle doves. *British Birds* 61: 193–212.

Murton R K, Westwood N J and Isaacson A J (1964) The feeding habits of the Woodpigeon *Columba palumbus*, Stock Dove *C. oenas* and Turtle Dove *Streptopelia turtur*. *Ibis* 106: 174–188.

2.4 Skylark

2.4.1 Habitat use in summer

Skylarks nest on the ground. Cereal crops, set-aside and grassland are the dominant habitats, although a wide range of arable crops are also used. They are particularly associated with vegetation 20–50 cm high. They avoid nesting close to hedges and even isolated pylons, trees or bushes in order to evade predators. Hence small, hedge-bound fields are unsuitable for skylarks. Semi-natural grasslands can hold the highest densities of breeding skylarks but, in the case of lowland farms, densities are generally highest on arable farms, intermediate on mixed farms and lowest on pastoral farms.

Vegetation structure plays an important role in determining breeding densities between differing crop types. On arable land, breeding densities are highest in spring-sown crops (except oilseed rape) and set-aside, and lowest in autumn-sown crops (especially oilseed rape). On grassland, silage fields attract breeding skylarks, but productivity is low due to nest loss during cutting. Breeding densities are low on improved grazed pasture.

Organic crops generally hold higher densities of breeding skylarks than conventional crops. There is also some evidence that breeding success is greater on organic farms. Differences are likely to be due to higher availability of invertebrate food, patchier and more open crops, which also allows more nesting attempts per season, and more diverse cropping on organic farms. In contrast, another study has shown than the tendency for organic fields to be smaller and more enclosed than on conventional farms, can result in no difference in numbers.

2.4.2 Habitat use in winter

Skylarks become communal in winter, flocking from late August to March in open fields away from field boundaries and generally not mixing with other species.

David Tipling/rspb-images.com

Plate 2.4 Skylark

Winter densities are highest in cereal stubble fields, harvested root crops, broad-leaved crops and rotational set-aside, and lowest in autumn-sown cereals and agricultural grasslands. Barley stubble provides the most important winter feeding habitat, probably because of the greater abundance of weed seeds. Stubbles with no pre- or post-harvest herbicide treatment are preferred. Rotational set-aside fields hold more skylarks in winter than non-rotational set-aside due to the greater abundance of seed food and easier access to the ground. Use of winter oilseed rape may be high in winters when crop establishment is poor as there is easy access to the ground and there may be a flush of broad-leaved weeds.

2.4.3 Diet

Skylarks eat seeds and plant leaves throughout the year and invertebrates in spring and summer. They prefer cereal grains and larger seeds (such as polygonums), but they do not take seeds less than 0.3 mg dry weight (such as speedwells, mayweeds and nettles).

Invertebrates are important in the adult diet from April until August and are the exclusive diet for chicks in their first few weeks. They take a wide variety of invertebrates, especially beetles and spiders. Chicks are fed soft-bodied invertebrates such as spiders, aphids and larvae for the first few days, with chitinous prey becoming more important as they grow. In the summer, they preferentially forage on set-aside, grass tracks, field margins, tramlines and other open areas (such as 'skylark plots') within arable crops and organic silage fields, whilst avoiding dense arable crops, improved grazed pasture and conventional silage. Preferences are probably due to a combination of invertebrate abundance and patchy vegetation cover that allows easy access to food.

The winter diet of skylarks on cereal stubbles is principally cereal grains, although the efficiency of modern harvesting machinery has reduced the availability of this food source. On set-aside, the principle food source is the leaves of broad-leaved weeds. Birds feeding on emerging cereal and broad-leaved crops also take green plant material. Leaves of broad-leaved plants are better than cereals and grasses because of their relatively high calorific value. Weed seeds are generally not a major component of the diet, probably due to the lack of availability. However, damage to emerging sugar beet crops in spring declines greatly as the availability of weed seeds and invertebrates increases, suggesting that the latter are more profitable food items.

2.4.4 Breeding

Skylarks breed in their first spring. More than half the adults that survive winter return to the territory used in the previous year – the likelihood increases if breeding was successful. Skylarks generally hold the same territory throughout the season and stop breeding activity when the vegetation becomes unsuitable for nesting.

Under suitable conditions, the breeding season lasts from early April to early August, with a peak of activity in May/June. Nests are built on the ground in crops, set-aside or grassland, preferentially where the vegetation height is 15–40 cm. Payments for extensive grazing in the South Downs Environmentally Sensitive Area (ESA), to maintain a sward above 15 cm high, led to an increase in skylark breeding densities. However, extensive grazing producing a sward shorter than 15 cm showed no benefits.

If habitat is available, skylarks can make up to four breeding attempts per season and normally lay four or five eggs per clutch; the peak clutch size is reached in May/June. Skylarks must have two to three successful broods per season for the UK population to be sustainable. Densities on farmland in Britain average 0.3 pairs per ha in set-aside, 0.1 pairs per ha in winter cereals, and as low as 0.05 pairs per ha in improved grassland.

A whole breeding attempt from nest building to the time that the young fledge takes 34–38 days for a typical brood of four chicks. However, because skylarks are unlikely to nest in newly cut silage fields, a minimum period of seven weeks between cuts is required to enable young to be successfully reared.

2.4.5 Population trend and distribution

Skylarks declined by 54% between 1970 and 2001. The sharpest decline occurred between the mid-1970s and the mid-1980s,

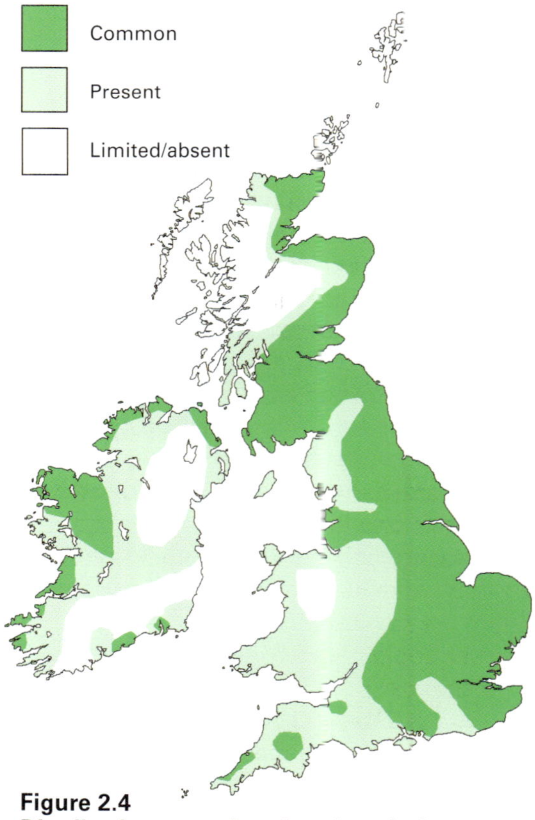

Figure 2.4
Distribution map showing the relative abundance of the skylark in Britain and Ireland in 1988–91.

with a more shallow decline since then. The most dramatic declines have occurred in the arable areas, where the highest densities occur (Figure 2.4). However, they remain widespread, having declined in range by only 2% in the UK between 1970 and 1990. The range contraction in Ireland was slightly higher (6%). Populations have also declined throughout mainland Europe.

2.4.6 Causes of decline

The breeding success of skylarks per breeding attempt has increased in recent years, and so can be disregarded as a cause of the recent decline. However, a decrease in the number of breeding attempts through the season could explain the decline. Population densities on farmland are related to the diversity of field types, suggesting that simplification of farm rotations may be one of the causes of decline. The main cause of decline on arable land is considered to be the widespread switch from spring to autumn-sown cereals. Autumn-sown cereals are taller and thicker throughout the breeding season, leading to:

- a reduction in territory densities
- a reduction in the number of nesting attempts made by each pair in a season
- a reduction in food availability
- a higher proportion of nests on or close to tramlines – leading to higher predation rates
- fewer winter stubble fields – the most important winter feeding habitat.

Experimental manipulation of winter cereal fields, by leaving two undrilled patches per hectare (each approximately 4m x 4m), increased skylark productivity by almost 50%. Skylarks nesting in fields with undrilled patches reared more broods per pair, fledged more chicks per brood and produced heavier chicks that were more likely to survive through the winter. If combined with measures to ensure overwinter survival, this scale of productivity would be sufficient to reverse the decline in arable areas if undrilled patches (or skylark plots) were widely adopted (in about 20% of winter cereal crops).

Changes in grassland management have also been implicated. The switch from hay to silage has increased nest failure rates caused by cutting, except in a few northern areas, such as the Hebrides, where the cutting date has not been brought forward significantly. In most of the UK, silage regimes damage skylarks because:

- the first cut is generally before late May – when most of the first skylark broods fledge
- the period between cuts is usually less than seven weeks – the minimum period needed for successful nesting

A further problem is increased stock densities, which create pasture swards too short for nesting skylarks. The additional risk of trampling is also increased as stock densities rise.

2.4.7 Management advice

Arable

Where possible, incorporate spring-sown cereals or break crops with a short, open structure (such as linseed) into the rotation. This will increase nesting densities, improve breeding success and prolong the nesting season. Where spring-sown cereals can be incorporated, a March sowing date provides ideal crop density and height for nesting skylarks throughout the breeding season. In general, a farm with a wider range of arable crops will present skylarks with more nesting and feeding opportunities than one with a more simple rotation.

An alternative is to create skylark plots (see section 3.17) (undrilled patches of 16–24 m^2) within winter cereal fields. These appear to open up the sward sufficiently to enable skylarks to forage freely within the field. Two patches per hectare work well in open fields of at least 5 ha. If tall hedges or woodland borders the field, then it should be at least 10 ha in size to be attractive to skylarks. The plots should not be connected to the tramlines or located within 24 m of the crop edge, to reduce the risk of predation.

Uncropped features, such as beetle banks (see section 3.10) within large, open fields away from field boundaries will boost insect-food abundance within the field, and may also be used by nesting skylarks.

The quality of feeding habitat can be improved by tolerating weed levels up to the economic threshold, where the savings from not applying herbicides balance with the associated yield losses. Broad-leaved weeds and cereal grains are excellent winter food for skylarks. This food can be made available by delaying the cultivation and spraying of overwinter stubbles for as long as possible. Where they can be left through the winter, stubbles provide the most important feeding habitat on the farm (see section 3.14).

Invertebrate food availability will be affected by pesticide use on the farm. Adoption of an integrated pesticide management policy should reduce pesticide use and benefit skylarks and a wide range of other farmland wildlife. Low-input crop management options (see section 3.13) under agri-environment schemes or organic crop management would boost benefits further.

On most arable farms, natural regeneration of rotational set-aside (see section 3.15) is the easiest way to provide weedy stubble for wintering skylarks. Whole-field set-aside, or 20 m strips through the middle of open fields, can provide good nesting habitat for skylarks. All forms of set-aside management should be delayed for as long as possible.

- Cutting destroys skylark nests – avoid cutting rotational set-aside and delay cutting fixed set-aside for as long as possible. Use the option to leave 25% uncut (for skylarks, ideally leave areas in the centre of the field, but be aware that other ground-nesting birds may be using field margins). The same area can be left uncut for up to three years, although it is best to rotate the uncut area around the field annually to prevent the vegetation becoming too rank for skylarks.
- Cultivation destroys skylark nests and prevents subsequent nesting attempts – delay cultivation of rotational set-aside for as long as possible.
- Spraying herbicides will not destroy

Table 2.4 Effects of different forms of land management on the skylark.

	Nesting	Seed food	Invertebrate food
Skylark plots (undrilled patches)	✔		✔
Beetle banks	?		✔
Non-rotational set-aside	✔		✔
Spring crops	✔		?
Mixed cropping	✔	✔	?
Rotational set-aside	?	✔	?
Arable pockets in pastoral areas	✔	✔	?
Low-input or organic crops			✔
Broad-spectrum pesticides		✘	✘
Hay meadows	✔	?	✔
Seven week gaps between silage cuts	✔		
Entensive grazing	?	?	✔
Overwintered stubble		✔	
Hedge and tree planting	✘	✘	✘

✔ indicates positive benefits
? indicates potential benefits dependent on management and outcome
✘ indicates negative impacts

nests or poison skylarks, but will leave nests more open to predation and discourage further nesting attempts – delay spraying for as long as possible.

Grassland

Cereal crops may be important refuges for skylarks within grass-dominated farmland. Use of spring cereals, or winter cereals with skylark plots, will boost skylark productivity (see section 3.18).

Skylarks will only breed successfully in silage fields if they are not grazed or cut between early April and the end of May, with subsequent cuts being at least seven weeks apart. It may be more practical to convert a proportion of the silage area to hay production, maintaining the remainder of the area as intensive silage (see section 3.19).

Extensive grazing will only provide suitable nesting habitat if the sward is maintained at over 15 cm (although shorter swards may provide useful feeding areas).

Where grassland is created for skylarks, use native grass species and do not use inorganic fertilisers. This will increase the invertebrate abundance and skylark breeding densities. These are conditions of grassland creation generally required by the agri-environment schemes (see section 3.20).

Table 2.4 shows beneficial options.

Further reading

Browne S J, Vickery J A and Chamberlain D E (2000) Densities and population estimates of breeding skylarks *Alauda arvensis* in Britain 1997. *Bird Study* 47: 52–65.

Chamberlain D E and Crick H Q P (1999) Population declines and reproductive performance of skylarks *Alauda arvensis* in different regions and habitats of the United Kingdom. *Ibis* 141 38–51.

Donald P F (2004) *The Skylark.* T & A D Poyser, London.

Donald P F, Buckingham D L, Moorcroft D, Muirhead L B, Evans A D and Kirby W B (2001) Habitat use and diet of skylarks *Alauda arvensis* wintering on lowland farmland in southern Britain. *Journal of Applied Ecology* 38: 536–547.

Donald P F and Vickery J A (2000) The importance of cereal fields to breeding and wintering skylarks Alauda arvensis in the UK. Pp 140–150 in N J Aebischer, A D Evans, P V Grice and JA Vickery (eds) *Ecology and Conservation of Lowland Farmland Birds.* British Ornithologists' Union, Tring.

Green R E (1978) Factors affecting the diet of farmland skylarks *Alauda arvensis. Journal of Animal Ecology* 47: 913–928.

Morris A J, Holland J M, Smith B and Jones N E (2004) Sustainable Arable Farming For an Improved Environment (SAFFIE): managing winter wheat sward structure for Skylarks *Alauda arvensis.* In Proceedings of The Ecology and Conservation of Lowland Farmland Birds II: The Road to Recover. *Ibis*: 146.

Odderskaer P, Prang A, Poulsen J G, Andersen P N and Elmegaard N (1997) Skylark (*Alauda arvensis*) utilisation of micro-habitats in spring barley fields. *Agriculture, Ecosystems and Environment* 62: 21–29.

Odderskaer P, Prang A, Elmegaard N and Andersen P N (1997) Skylark reproduction in pesticide treated and untreated fields. *Pesticides Research* 32, Danish Environmental Protection Agency, Copenhagen.

Wakeham-Dawson A, Szoszkiewicz K, Stern K and Aebischer N J (1998) Breeding skylarks *Alauda arvensis* on Environmentally Sensitive Area arable reversion grass in southern England: survey-based and experimental determination of density. *Journal of Applied Ecology* 35: 635–648.

Wilson J D, Evans J, Browne S J and King J R (1997) Territory distribution and breeding success of skylarks *Alauda arvensis* on organic and intensive farmland in southern England. *Journal of Applied Ecology* 34: 1462–1478.

Wilson J D, Taylor R and Muirhead L B (1996) Field use by farmland birds in winter: an analysis of field type preferences using re-sampling methods. *Bird Study* 43: 320–332.

2.5 Tree sparrow

2.5.1 Habitat use in summer

The tree sparrow breeds on lowland farmland and is particularly associated with coastal areas, mixed farming areas, farmland alongside slow-flowing, lowland watercourses and water-bodies, parkland and small woods within farmland. Despite the species' name, larger woodlands are usually avoided, only being occupied when population density in more favoured open countryside habitats is at a maximum. Tree sparrows occasionally nest in suburbs, but within the UK are largely replaced by house sparrows in areas of human habitation. They are a colonial nesting species, with colonies generally consisting of up to 50 breeding pairs.

2.5.2 Habitat use in winter

Birds evacuate the nesting colony once breeding has finished. Post-breeding flocks form around cereal fields and waste ground where there is an abundance of weed seeds and they roost in thick hedge cover. Several colonies may combine to form larger wintering flocks that range up to 5–10 km from the breeding areas. Adults may return to the breeding colonies after moulting in late September/October to roost, depending on winter foraging potential around the nesting area. This may be to defend nesting holes, which can be a limited resource in some areas. Otherwise, birds return in March/April. In mid-winter, seed-rich areas such as stubble fields, weedy areas and wild bird covers are used for foraging. Tree sparrows may be found in mixed flocks with finches and buntings.

Andy Hay/rspb-images.com

Plate 2.5
Tree sparrow

Large-scale autumn eruptive movements from the north of the range occur sporadically, especially by first year birds. The last recorded in the UK occurred between 1957 and 1962. These movements may be a response to high population densities in the birds' natal areas.

2.5.3 Diet

Adults feed largely on plant matter. This forms virtually 100% of their diet between October and March. During the summer, a variable proportion of insect food is included. Important components of the adult diet are the seeds of cereals, polygonums (eg knotgrass and redshank) and fat hen. Other regular components include chickweed, shepherd's purse, forget-me-not and grass weeds but the precise diet composition in any situation is likely to vary greatly according to local availability. Weed seeds are particularly important in the first half of the year, and cereals dominate the diet from the point when the crops reach the milky-ripe stage in mid-summer through until the end of the year. Cereal grain generally makes up to one-third of their diet from July to September and up to half of their diet between October and December. They will also eat a wide variety of commercial bird foods, including sunflower seeds, millet and peanuts, and other crops such as rape seed and crushed maize. They also eat buds, shoots, berries and flowers from fruit trees. Seeds are largely collected on the ground.

The chicks are fed largely on insects and spiders (more than 80% of diet by dry weight) until they fledge, when they switch to a seed-based diet. Composition of the invertebrate diet is largely opportunistic, but spiders, beetles and soft-bodied larvae are consistently important. Flies and bugs (including aphids) are also taken regularly and populations close to water bodies often include a high proportion of aquatic insects, such as midges and damselflies, in the diet. Invertebrates are usually taken from the ground or gleaned from vegetation including crops, grass field margins, trees, hedgerows, pasture and around water bodies. Chick food is usually collected within 500 m of the nest site.

2.5.4 Breeding

Tree sparrows generally nest in holes in trees, mature hedges, farm buildings or nestboxes. They rarely build free-standing nests in scrub or buildings. The nest is usually between 2 m and 8 m high. Adults tend to be faithful to the nesting area that they used in previous years. First year birds tend to move to other breeding colonies occupied by birds that they joined in the wintering flocks. There can be both intra- and inter-specific competition for nesting holes. Tree sparrows can dispossess tits and other small, hole-nesting birds from nest sites, and can in turn be dispossessed by house sparrows.

Egg-laying occurs mainly between mid-April and the end of July. Tree sparrows in Europe make up to three (very rarely four) breeding attempts. A British study estimated that roughly 50% of the population had a single clutch in the year, 40% had two clutches and 10% had three clutches. The distribution of egg-laying dates shows three peaks in early May, mid-June and mid- to late July, which relate to first, second and third broods respectively. The main periods of foraging for chick-food insects are therefore late May for the first brood, late June/early July for the second brood, and early August for the third brood.

Clutch size can vary from one to eight eggs (most commonly four to six). Incubation takes 11–14 days and the chicks fledge after 15–20 days, but are still fed by the adults for a further 10–14 days.

Nesting failure can be the result of infertile eggs or predation. Chick mortality can be due to predation, starvation or chilling in periods of cold weather.

2.5.5 Population trend and distribution

Historically, there is evidence that the tree sparrow population has fluctuated widely within the UK. Data compiled by the BTO indicates that there was a six-fold increase in the UK population between the mid-1950s and the mid-1960s. The species recolonised Ireland during this period, having become extinct in the early 1950s. This increase was followed by a population crash from the mid-1970s in line with a range of other seed-eating farmland birds. Common Bird Census data indicate a 94% decline in numbers from 1970 to 2001. The more recent Breeding Bird Survey suggests that the decline may have stabilised in the UK since the early 1990s and that some recent regional increases may have taken place. It is possible that the current UK population is still as low as, or even lower than, the previous low numbers reached in the 1950s. Figure 2.5 shows the relative abundance of tree sparrows in 1988–91. Since then, there have been range

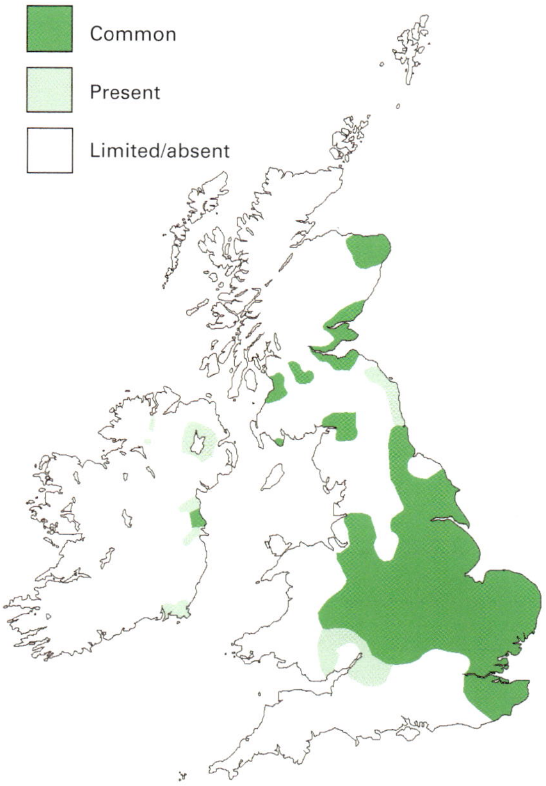

Figure 2.5
Distribution map showing the relative abundance of the tree sparrow in Britain and Ireland in 1988–91.

contractions, especially in East Anglia and south-east England.

2.5.6 Causes of decline

Breeding success per nesting attempt has increased during the population decline, but it is unknown whether or not the number of nesting attempts per pair has changed. Ringing recoveries show that there is a high rate of adult mortality during the nesting period (April to July), presumably due to the stress imposed on them through rearing young, either through the physical effort or because they are put under greater predation risk when foraging for food.

A research project at Rutland Water has shown that tree sparrows will feed a wide variety of invertebrates to their chicks. The study showed the importance of aquatic invertebrates (midges, damselflies etc.) in chick diet in some years, indicating that wetlands may provide important food resources when insect abundance on farmland is low. A clear preference was shown for colonising new clusters of nestboxes installed within 1 km of the wate body. This may explain why the current distribution of tree sparrows often coincides with water bodies and river valleys.

Availability of suitable nesting holes may be a limiting factor for this colonial nesting species in some areas.

2.5.7 Management advice

Arable

Tree sparrows forage for insects readily in rough grass field margins of arable fields (see section 3.9). Select margins that have thick hedges for cover. They will also benefit from measures to boost invertebrate abundance within the crop. Only use insecticides when pest infestations exceed the economic thresholds. Try to avoid using broad-spectrum insecticides after 15 March. These remove beneficial insects and spiders that move into the crops in the spring. The loss of this food source may be particularly damaging to tree sparrows. Conservation headlands (see section 3.12) (limited pesticide use in cropped headlands) and low-input crops are useful methods for boosting insect food availability in the spring. Alternatively, broad-leaved weeds and their associated insects can be encouraged using cultivated uncropped margins.

Spray and cultivate overwinter stubbles (see section 3.14) as late as possible. This provides important winter-feeding habitat. Natural regeneration on rotational set-aside provides the same habitat where overwinter stubbles preceding spring crops are not a viable option. Weedy stubbles are more valuable – for this reason, stubbles are better if they do not receive any pre- or post-harvest herbicide. Alternatively, you can provide seed food throughout the winter by creating areas of wild bird cover (see section 3.7). Kale and quinoa are particularly useful components in the mix, although tree sparrows take a wide range of seed types.

Grassland

Introduce arable fodder crops (see section 3.18) or small plots of wild bird cover (see section 3.7) to provide a seed-rich habitat in pastoral areas. Maize is generally not of value to tree sparrows due to the low weed numbers present, although maize grain in the stubble and silage may be eaten. Alternatively, some seed food can be provided by fencing off margins of up to 6 m around improved grassland and leave these unfertilised, uncut and ungrazed. Graze or cut in September every two to three years. Select margins that have thick hedges for cover (see section 3.24).

Protecting nest sites

You could ask a local birdwatcher or your county bird club to find the nesting area of your tree sparrow colony. This will ensure the colony is safe from accidental damage. Maintain large, thick hedges on the farm and retain any old bushes, trees or farm buildings that may contain nesting holes (see sections 3.1, 3.3 and

Table 2.5 Effects of different forms of land management on the tree sparrow.

	Nesting	Seed food	Insect food
Maintain mature hedges and trees	✔		
Farm buildings	✔		
Nestboxes	✔		
Overwintered stubble		✔	
Wild bird cover		✔	?
Rotational set-aside		✔	?
Arable pockets in pastoral areas		✔	?
Conservation headlands		?	✔
Low-input cereal		?	✔
Broad-spectrum pesticides		✘	✘
Rough grass field margins		?	✔
Cultivated margins		✔	✔
Wetland features			✔

✔ indicates positive benefits
? indicates potential benefits dependent on management and outcome
✘ indicates negative impacts

3.5). Continue to manage old pollard trees and re-pollard neglected trees, as these provide nest sites (see section 3.2). Nestboxes can be used to supplement the number of nest sites. Boxes should have 2.8 cm diameter entrance holes and an internal box depth of 20 cm. Tree sparrows nest in colonies, so place several boxes on each tree, close together and above head height

Wetland features

Insect abundance may be boosted by the creation of ponds or the installation of sluices in ditches to create open water features throughout the summer (see section 3.4).

Table 2.5 summarises the beneficial options.

Further reading

Anderson G Q A, Gruar D J, Wilkinson N I and Field, R H (2002) Tree sparrow *Passer montanus* chick diet and productivity in an expanding colony. *Aspects of Applied Biology* 67: 35–42.

Field R H and Anderson G Q A (2004) Habitat use by breeding tree sparrows *Passer montanus*. *Ibis* 146: 60–68.

Perkins A J and Anderson G Q A (2002) Seed selection by tree sparrows *Passer montanus*: determining appropriate seeds for supplementary feeding on farmland. *Aspects of Applied Biology* 67: 213–220.

Summers-Smith J D (1995) *The Tree Sparrow*. J Denis Summers-Smith, Guisborough, UK.

2.6 Linnet

2.6.1 Habitat use in summer

The linnet is typically a bird of farmland and waste ground and found wherever there are abundant sources of seeds. In the UK, 70% of linnets nest on farmland. They nest solitarily or in loose colonies low down in a dense, thorny hedge or bush (especially gorse early in the season). Breeding densities are higher in areas with smaller field sizes and a higher proportion of fallow land. Where there is a shortage of broad-leaved weed seeds, foraging is targeted on grasslands with dandelion (including road verges) early in the season and oilseed rape fields (where these are available) from the point when the seeds are ripening in the pods.

2.6.2 Habitat use in winter

Linnets form large flocks, often with other seed eaters, in order to find new patches of food plants and to detect predators. A study in East Anglia found that 80% of the wintering population was on set-aside, whilst only 1% was found on winter cereals. Weedy stubbles are the favoured feeding habitat, but fields which develop a lush growth of cereal volunteers are not as attractive to linnets or other seed eaters, as access to seeds on the ground is reduced. Likewise, undersown stubbles reduce the accessibility of surface seed and the growth of the grass suppresses germination of broad-leaved weeds. Linnets select areas with high densities of accessible seed (over 250/m^2) and a greater proportion of bare earth.

Variable numbers of birds move south, or migrate to Europe or Ireland in winter.

2.6.3 Diet

Linnets feed communally in areas with abundant food, and travel up to 2 km from the nest to feed. Unlike many of the other declining seed-eating birds, linnets also feed their chicks on seeds. They eat small to medium-sized seeds (less than 10 mg) including most arable weeds, especially polygonums (eg knotgrass, docks), crucifers (eg charlock, shepherd's purse), chickweeds and composites (eg dandelions, thistles, sow-thistles, mayweeds and groundsel). They prefer milky seeds, and ripe ones are often avoided. Some seeds and fruit of trees and shrubs are occasionally taken (eg hawthorn and birch). There is a small component of invertebrates in the diet. The importance of this is unknown. Linnets feed on the ground, on herbaceous vegetation and low down in bushes.

There has been a marked shift in the diet of linnets in Oxfordshire from a wide

Chris Gomersall/rspb-images.com

Gerald Downey/rspb-images.com

Plate 2.6 (left)
Male linnet
Plate 2.7 (right)
Female linnet

diversity of broad-leaved weed seeds in the early 1960s to a dominance of dandelion and oilseed rape in the late 1990s.

2.6.4 Breeding

Linnets breed in the first spring after fledging. They nest in dense, thorny hedgerows or areas of scrub. Hawthorn, bramble and blackthorn are the three most common shrubs used. They nest in a wide variety of hedgerow types but preferentially select hedgerows of 1–2 m.

There are usually two breeding attempts, with three in favourable conditions. The eggs are laid between mid-April and early August in southern England with a peak in late April to early May for the first brood and early June for the second brood. The last broods can fledge in early September. The clutch size is usually four to six eggs (rarely three to seven) and generally increases from April to June. The breeding cycle consists of nest building (four to five days), egg-laying (four to six days), incubation (11–13 days), and the chicks fledge after 10–17 days so the full cycle is four to six weeks. The young are fed up to two weeks after fledging. There are reports of overlapping broods where the male feeds fledged chicks whilst the female starts to incubate a new clutch of eggs.

2.6.5 Population trend and distribution

The UK linnet population declined by 51% between 1970 and 2001. The main decline occurred between 1975 and 1986, and the population has remained relatively stable since then, although with fluctuations. This species remains widespread, especially in lowland areas. Figure 2.6 shows the relative abundance of linnets in 1988–91.

2.6.6 Causes of decline

An increase in the nest failure rate is the most likely demographic mechanism that drove the decline of this species. The decline was probably the result of reduced availability of weed seeds in both arable and pastoral farmland. On arable land, this is largely due to the increased use and efficacy of herbicides. The diet of linnet chicks in Oxfordshire in the 1960s was based on a wider range of weed seeds than the diet in the 1990s, when broods were fed predominantly on dandelion in May, and on oilseed rape between June and August.

The decline largely occurred before the widespread production of oilseed rape and probably arose through a reduction in the number of chicks fledged per nest and a reduction in the weight of chicks at fledging, which could lead to increased post-fledging mortality. An increase in the area of oilseed rape, and the development of more palatable varieties, led to an increase in breeding success in the 1990s, resulting in higher breeding densities, lower nest failure rates, lower chick starvation rates and an earlier start to the breeding season. This increased the chances of birds being able to make three breeding attempts. The distance from the nest to the nearest oilseed rape field significantly influenced the growth rate of

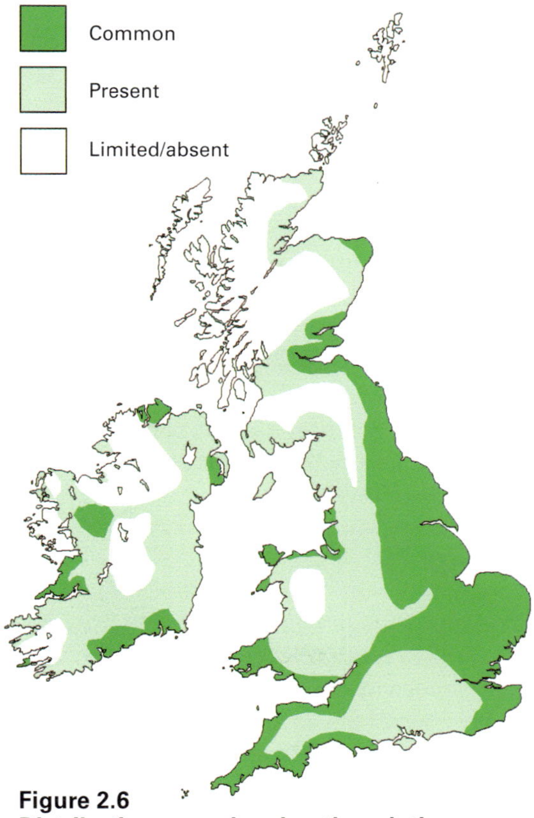

Figure 2.6
Distribution map showing the relative abundance of the linnet in Britain and Ireland in 1988–91.

chicks in the nest, which in turn significantly reduced the probability of chick starvation. Overall, breeding productivity on farmland where oilseed rape is available is high enough to sustain the local population if females are capable of attempting three broods in a season. In a study area in Norfolk, where oilseed rape was not present in the rotation in the late 1990s, the diet was predominately dandelion and grass seeds. The rate of chick starvation was higher in this area than in an area of Oxfordshire where oilseed rape was abundant. Future declines are therefore likely in areas where oilseed rape is not grown, or if the area of oilseed rape declines.

Another factor which may have assisted this recovery was the introduction of set-aside, on which a large proportion of the linnet population feeds through the winter. However, linnet survival measured by ring-recovery data has not significantly altered over this period.

Food availability seems to override the effect of nest-site availability in limiting breeding populations of linnets, despite the highly publicised loss of hedgerows before 1990. However, the reduction in the quantity and quality of hedgerows as cover for nesting attempts may be an additional factor in some areas.

2.6.7 Management advice

General

Any verges and areas of waste ground can provide important seed food for linnets if dandelion or any annual plants are allowed to seed.

Arable

The important resource for linnets is year-round seed food. This can be provided using overwinter stubbles (see section 3.14) or wild bird cover (see section 3.7) (wild bird seed mixture or unharvested crops) in the winter, and cultivated field margins, low-imput cereals or conservation headlands in the summer (see sections 3.11–3.13).

Delay spraying and cultivation of stubbles for as long as possible, preferably until March. Natural regeneration on rotational set-aside (see section 3.15) is a useful option where stubbles are not otherwise possible in the arable rotation. This can provide food throughout the year if spraying and cultivation can be delayed to provide benefits through the breeding season.

Alternatively, creation of small areas of wild bird cover incorporating species that produce small, oil-rich seeds, such as kale, quinoa, mustard or oilseed rape can provide an abundant seed source.

Mixed rotations which incorporate spring and winter oilseed rape will provide seed food throughout much of the summer, but this can be supplemented using conservation headlands in cereal crops or cultivated, uncropped margins. These options are more likely to be acceptable on light soils where the weed burden is less likely to be dominated by competitive grass weeds and cleavers. In general, only using herbicides when weed numbers exceed the economic threshold will improve the feeding habitat quality throughout the year.

Unmanaged field margins, field corners and ditch banks will provide grass seed. This is not a favoured food for linnets but will be utilised if availability of other seed sources is limited (see section 3.9 and section 3.4).

Grassland

In farmland areas dominated by grassland, the main benefit would be through retention or creation of arable pockets using arable fodder crops (see section 3.18) for stock feed, particularly root crops such as turnips where the weed flora is allowed to develop.

Restoration or creation of hay meadows can provide important seed sources in pastoral areas, especially if seed from species-rich grasslands can be incorporated (see section 3.19).

Keeping 6 m margins around grass fields uncut and ungrazed through the summer

Table 2.6 Effects of different forms of land management on the linnet.

	Nesting	Summer seed food	Winter seed food
Hedgerow management	✔		
Maintain areas of scrub and bramble	✔		
Arable pockets in pastoral areas		✔	✔
Wild bird cover		?	✔
Rotational set-aside		?	✔
Mixed cropping			✔
Overwintered stubble			✔
Cultivated margins		✔	?
Conservation headlands		✔	?
Low-input cereals		✔	?
Broad-spectrum herbicides		✘	✘
Unmanaged verges and waste ground		✔	✔
Hay meadows		✔	✔
Rough grass field and ditch margins			?

✔ indicates positive benefits
? indicates potential benefits dependent on management and outcome
✘ indicates negative impacts

and not applying fertiliser may create a grass sward rich in seeds. Linnets may use these in the absence of sources of smaller seeds (see section 3.24).

Hedgerows

Maintaining short, thick hedgerows (see section 3.1) or areas of scrub (see section 3.3) made up of thorn species, gorse or bramble will provide ideal nesting habitat for linnets.

Table 2.6 summarises the beneficial options.

Further reading

Eybert M C, Constant P and Lefeuvre J C (1995) Effects of changes in agricultural landscape on a breeding population of linnets, *Acanthis cannabina L.* living in adjacent heathland. *Biological Conservation* 74: 195–202.

Gregory R D and Ballie S R (1998) Large-scale habitat use of some declining British birds. *Journal of Applied Ecology* 35: 785–799.

Moorcroft D, Bradbury R B and Wilson J D (1997). The diet of nestling linnets, *Carduelis cannabina*, before and after agricultural intensification. Vol 3, pp 923–928 in R C Kirkwood (ed) *The 1997 Brighton Crop Protection Conference – Weeds*. British Crop Protection Council, Brighton.

Moorcroft D and Wilson J D (2000). The ecology of linnets, *Carduelis cannabina*, on lowland farmland. In N J Aebischer, A D Evans and J Vickery (eds) *Proceedings of the 1999 BOU Spring Conference, Lowland Farmland Birds*. British Ornithologists' Union, Tring.

Wilson J D, Taylor R and Muirhead L B (1996). Field use by farmland birds in winter: an analysis of field type preferences using re-sampling methods. *Bird Study* 43: 320–332.

2.7 Yellowhammer

2.7.1 Habitat use in summer

The yellowhammer is found on a wide range of farmland types, but is most common on lowland arable and mixed farmland. Yellowhammers nest along hedges, in patches of scrub or ditches and are particularly associated with boundaries that have wide grass margins. Short hedges (up to 2 m) are preferred. The early nests, before the hedge is in full leaf, tend to be in the herbaceous vegetation of the ditch or field margin abutting the hedge, with later nests in the hedge itself.

During the summer, they forage close to the nest site by hedges, on farm tracks and waste ground, and in open fields. Cereal fields become particularly important foraging sites when the crops come into ear and birds can eat the ripening grain. High breeding densities are associated with areas that have a high proportion of cereal area, mixed arable cropping and a high density of hedgerow boundaries. Low densities are found on pastoral farmland and areas with few cereal crops. When foraging for chick food, selected habitats include field margin habitats, cereal crops, natural regeneration on non-rotational set-aside and unimproved grasslands. Habitats that are consistently under-used include improved grasslands, woodlands, watercourses and rotational set-aside fields that are sprayed with glyphosate before yellowhammers have chicks in the nest (mid-May onwards). A sparse sward structure is an important feature of foraging habitat, allowing access to the ground.

2.7.2 Habitat use in winter

About 70% of adult yellowhammers in the UK winter within 5 km of their breeding territories. There is a tendency for birds to retreat from the uplands and there is a more marked seasonal movement of birds in Ireland. Flocks start to form in late July and break up in March. Yellowhammers feed in flocks with other seed eaters at good feeding sites throughout the winter, such as stubbles, newly sown fields, cereal-based game cover, grain in farmyards and at stock and game-feeding sites.

A research project in Oxfordshire found that wintering yellowhammers were found in significantly greater numbers than expected in winter stubble fields, and in significantly fewer numbers than expected in grazed grassland, conventional winter cereals and broad-leaved crops. Numbers on ungrazed grassland, bare tillage and organic winter cereals did not differ significantly from a random distribution.

Roger Wilmshurst/rspb-images.com

Tony Hamblin/rspb-images.com

Plate 2.8 (left) Male yellowhammer Plate 2.9 (right) Female yellowhammer

2.7.3 Diet

Adults feed on seeds, especially of cereals and other grasses (eg meadow grasses, fescues and ryegrass). Other plant matter includes seeds of weeds such as nettle, dock, knotgrass, fat hen, chickweed, mouse-ear, bramble, vetches, clover, forget-me-not, dandelion, knapweed, sow-thistle, yarrow and plantains. They seem to avoid oily seeds such as those of crucifers.

Invertebrates are taken in the breeding season and, to a lesser extent, throughout the remainder of the year. These include springtails, mayflies, grasshoppers, cockroaches, earwigs, bugs, lacewings, caddis flies, sawflies, spiders, woodlice, millipedes, caterpillars, flies, beetles, earthworms and snails. They are more able to feed on slow-moving invertebrates. The chicks are fed predominantly on invertebrates, but also on unripe cereal grain. Cereal grain was relatively rare or absent in the nestling diet up until the middle of the 20th century. It is possible that grain is sub-optimal food for chicks, and used only when insufficient insects are available. Yellowhammers generally feed on the ground, although some insects are taken from hedgerows.

2.7.4 Breeding

Yellowhammers start breeding in the first summer after hatching. They are monogamous. Territories tend to be linear, along hedgerows. Males start defending territories from early February. Egg-laying starts in late April or early May in southern England, and a little later further north. The egg-laying period extends until the end of July (peaks in May–June). Yellowhammers can make up to three nesting attempts per year, although they can only rear a maximum of two broods (third attempts are only made if one of the earlier attempts failed). The nest is built on or close to the ground, hidden in herbage at the base of a bank or hedge or within bramble. Ground nests are more common in April than later in the season. The breeding cycle consists of:

- nest building (four to five days by the female)
- egg-laying (three to five days)
- Incubation (12–14 days by the female)
- chicks reared to fledging (11–13 days)
- fledged chicks continue to be fed for a further eight to 14 days.

As a result, late nests can continue to be active throughout August. The female may start a second nest whist the male feeds the first brood. The clutch generally consists of three to five eggs. The majority of nest failures are due to predation. Nests in ditches suffer a higher predation rate than nests in hedges. Nests may be predated by crows or small rodents. The survival of broods increases through the season.

2.7.5 Population trend and distribution

The yellowhammer is the commonest and most widespread bunting in Europe. Whilst still widespread in the UK, the yellowhammer is scarce in pastoral areas and there were contractions in the range during the 1970s and 1980s, particularly in

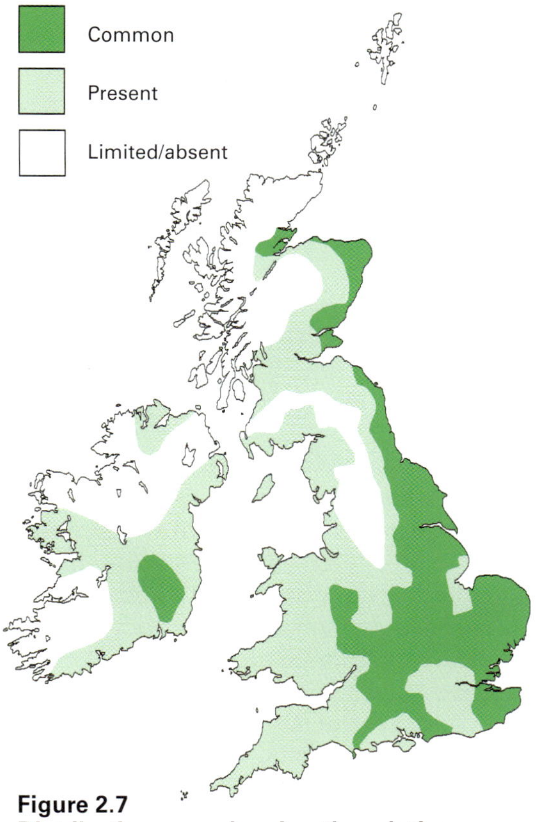

Figure 2.7
Distribution map showing the relative abundance of the yellowhammer in Britain and Ireland in 1988–91.

Ireland, but also in Wales, western Scotland and the upland fringes of northern England (Figure 2.7). Unlike many of the declining farmland species that showed marked declines between the mid-1970s and mid-1980s, the yellowhammer was relatively stable through this period, and declined more seriously in the late 1980s and early 1990s. The overall effect in the UK was a decline of 52% between 1970 and 2001.

2.7.6 Causes of decline

While there is some evidence that survival rates have decreased during the period of decline, yellowhammer breeding performance has tended to increase. However, recent declines in brood size and egg-stage failure rates are potentially of concern. Later nests tend to be more successful. The average number of broods per pair may have declined. However, the main cause of the decline is probably a decline in post-fledging survival and overwinter survival that is related to low availability of seed-rich habitats on farmland in the UK.

Yellowhammer nesting densities are greatest along hedgerows with wide grass margins in arable fields. If winter habitat is not the most limiting resource, then availability of this type of field boundary may limit nesting densities in some areas.

2.7.7 Management advice

Arable

Traditionally, overwinter stubbles (see section 3.14) were the most important winter feeding habitat on an arable farm. Yellowhammers fed with other seed-eating birds on spilt grain and weed seeds. Clean stubbles are therefore less beneficial than weedy stubbles. Stubbles following conservation headland management (see section 3.12) or a low-input cereal (see section 3.13) are ideal. Any use of herbicides pre-harvest or on the stubbles will reduce the food provided. Selective herbicides against the target weeds are better than broad-spectrum herbicides. Spray and cultivate stubbles as late as possible. Rotational set-aside (see section 3.15) should not be sprayed until the start of seeding of the target grass weeds to gain good control and maximum benefit for yellowhammers. In southern England, this is likely to be around the end of May for black grass and barren brome, early June for wild oats and July for couch grass. Timings will be later further north, but will always be best timed by observation of the growth stage of the weeds. Recent research has shown that topping the stubble (from 12 cm to 3 cm) can increase access to food for species such as yellowhammer.

Alternatively, wild bird cover (see section 3.7) can be established in the spring to provide a winter seed source. A high cereal proportion will be particularly beneficial for yellowhammers. Triticale is the cereal with the best standing power through the winter. Wild bird cover mixes can either be replaced every year, using a mix of annual seed-producing plants such as cereals, quinoa, oilseed rape, mustard or linseed, etc, or biennial, using kale as the brassica component.

Rough grass margins adjacent to short, thick hedgerows are beneficial in arable and grass fields. Margins of 6 m or more are ideal, although 1–2 m margins are better than ploughing up to the hedge-base. The more variable the grass sward is, the better to provide overwintering refuges for insects and suitable foraging habitat for yellowhammers. This can be created by a mix of grasses, including tussock-forming species such as cocksfoot, and fine, low-growing species such as fescues and bents. Variety can also be enhanced by different cutting regimes: long rotations will encourage tussocky margins, whilst regular cutting will provide more accessible ground for foraging. A wide margin where the strip next to the hedge is allowed to develop into a tussocky sward and the strip next to the crop is cut short (outside of the nesting season) is probably ideal (see section 3.9).

Grassland

Yellowhammers are liable to local extinction in purely pastoral farmland areas, where the shortage of seed food is the limiting factor. Arable pockets in pastoral areas are known to be very important for yellowhammers. Cereal crops in particular provide them with food in the summer and, if left as overwinter stubble, through the winter too. The same stubble management as explained above will apply. Whole-crop silage will be less beneficial as no grain will be dropped at harvest, although some food provision in the form of weed seed will still occur (see section 3.18).

Nesting habitat on pastoral farmland will also benefit from having rough grass margins adjacent to suitable hedgerows for nesting. They will benefit from margins that remain uncut and ungrazed for several years and receive no fertiliser inputs so that a relatively diverse tussocky sward develops which holds a greater abundance of invertebrates (see section 3.24). However, these areas will be marginal nesting habitats for yellowhammers as they will prefer locations close to relatively rich seed sources such as cereal fields. Recent research has found that silage fields that are normally cut three times a year, but are left to go to seed after the second cut attract large numbers of yellowhammers.

Hedgerow management

Manage hedgerows (see section 3.1) to maintain a continuous thick base by rotational trimming. Thorny hedgerows up to 2 m tall are ideal. If hedgerows are becoming tall and leggy, or gaps are appearing, then restoration by laying or coppicing may be necessary. Only restore short sections each year to retain plenty of cover throughout this period. Wide, tussocky grass margins adjacent to the hedge are important for nesting and to build up populations of insects for food during the breeding season. If sections of these margins are trimmed annually, then the shorter vegetation may provide more suitable foraging habitat for the yellowhammer. For example, a 6 m margin could have the hedge-side 3 m cut once every three years and the crop-side 3 m cut annually to provide a mix of insect-rich habitat and foraging habitat. Avoid hedgerow, ditch and field margin management between April and the end of September.

Table 2.7 summarises the beneficial options.

Further reading

Bradbury R B, Kyrkos A, Morris A J, Clark S C, Perkins A J and Wilson J D (2000) Habitat associations and breeding success of yellowhammers on lowland farmland. *Journal of Applied Ecology* 37: 789–805.

Table 2.7 Effects of different forms of land management on the yellowhammer.

	Nesting	Seed food	Insect food
Hedgerow management	✔		
20 metre non-rotational set-aside strips	✔		✔
Rough grass field margins and ditch banks	✔	?	✔
Arable pockets in pastoral areas		✔	?
Wild bird cover		✔	?
Rotational set-aside		✔	?
Overwintered stubble		✔	
Conservation headlands		?	✔
Low-input spring cereals		?	✔
Broad-spectrum herbicides		✘	✘
Cultivated margins			?

✔ indicates positive benefits
? indicates potential benefits dependent on management and outcome
✘ indicates negative impacts

Bradbury R B and Stoate C (2000) The ecology of yellowhammers on lowland farmland. Pp 165–172 in N J Aebischer, A D Evans, P V Grice and J A Vickery (eds) *Ecology and Conservation of Lowland Farmland Birds*. British Ornithologists' Union, Tring.

Morris A J, Bradbury R B and Wilson J D (2002) Determinants of patch selection by yellowhammers *Emberiza citrinella* foraging in cereal crops. *Aspects of Applied Biology* 67: 43–50.

Morris A J, Whittingham M J, Bradbury R B, Wilson J D, Kyrkos A, Buckingham D L and Evans A D (2001) Foraging habitat selection by yellowhammers (*Emberiza citrinella*) nesting in agriculturally contrasting regions in lowland England. *Biological Conservation* 101: 197–210.

Perkins A J, Whittingham M J, Morris A J, Barnett P R, Wilson J D and Bradbury R B (2000) Habitat characteristics affecting use of lowland agricultural grassland by birds. *Biological Conservation* 95: 279–294.

Perkins A J, Whittingham M J, Morris A J and Bradbury R B (2002) Use of field margins by foraging yellowhammers *Emberiza citrinella*. *Agriculture, Ecosystems and Environment* 93: 413–420.

Siriwardena G M, Baillie S R, Crick H Q P, Wilson J D and Gates S (2000) The demography of lowland farmland birds. Pp 117–133 in N J Aebischer, A D Evans, P V Grice and J A Vickery (eds) *Ecology and Conservation of Lowland Farmland Birds*. British Ornithologists' Union, Tring.

Siriwardena G M, Baillie S R and Wilson J D (1998) Variation in the survival rates of some British passerines with respect to their population trends on farmland. *Bird Study* 45: 276–292.

Stoate C (1999) The influence of field boundary structure on breeding territory establishment of whitethroat *Sylvia communis* and yellowhammer *Emberiza citrinella*. *Aspects of Applied Biology* 54: 125–130.

Stoate C, Moreby S J and Szczur J (1998) Breeding ecology of farmland yellowhammers *Emberiza citrinella*. *Bird Study* 45: 109–121.

Stoate C and Szczur J (2001) Whitethroat *Sylvia communis* and yellowhammer *Emberiza citrinella* nesting success and breeding distribution in relation to field boundary vegetation. *Bird Study* 48: 229–235.

Wilson J D, Taylor R and Muirhead L B (1996) Field use by farmland birds in winter: an analysis of field type preferences using re-sampling methods. *Bird Study* 43: 320–332.

2.8 Reed bunting

2.8.1 Habitat use in summer

The reed bunting is often found in wetland habitats, along riversides and close to water bodies, but this association is probably linked to the requirement for tall, thick vegetation, rather than an association with water. In arable situations, it is often found nesting in oilseed rape crops away from water, and it is found in good numbers in rank grassland habitats. Cereals and set-aside tend only to be used if close to wet features, and densities in oilseed rape crops were about four times greater in a study in Nottinghamshire. Use of broad-spectrum herbicides on rotational set-aside in late-April or May is likely to reduce the value of these fields for both nesting and foraging. Abundance of insect food for the chicks is probably a key factor influencing habitat choice. Rank vegetation also provides better concealment for nests from predation, and nest survival is directly related to the extent of nest concealment.

2.8.2 Habitat use in winter

Apart from a movement away from upland areas in the winter, the reed bunting is largely sedentary in the UK. About 40% move less than 5 km between summer and winter areas. They roost in thick hedges and feed in seed-rich areas of farmland, such as set-aside, overwintered stubbles and wild bird cover.

2.8.3 Diet

The reed bunting is mainly a seed-eating bird with insects taken during the breeding season, and opportunistically throughout the rest of the year. The important seed groups in the diet include brassicas, grasses and fat hen. Important invertebrate groups include spiders, caterpillars, grasshoppers, beetles and aquatic insects. Spiders and caterpillars formed 70% of the insect diet of chicks in a Nottinghamshire study, but it is likely that the composition varies with relative availability of the key insect groups.

2.8.4 Breeding

The nest is well hidden in thick vegetation, usually on the ground, but reed buntings can also nest in bushes. Eggs are laid from late April until mid-July. The clutch is typically of four to five eggs, but may vary from three to seven eggs. Reed buntings generally rear one or two broods per year. Incubation takes 12–15 days and the chicks fledge after nine to 13 days. The period from egg-laying to fledging is therefore about four weeks, although chicks may leave the nest three to five days before fledging. Parents continue to tend to young for about 20 days after they leave the nest. Nest predators include carrion crow, magpie, weasel, fox and rat.

Mike Richards/rspb-images.com

Bob Glover/rspb-images.com

Plate 2.10 (left) Male reed bunting
Plate 2.11 (right) Female reed bunting

2.8.5 Population trend and distribution

The reed bunting is widely distributed in Britain and Ireland, though densities are generally higher in Ireland than Britain. The population in the UK increased between 1963 and 1975, probably as a result of a gradual move into drier habitats such as arable farmland. There was a subsequent sharp decline between 1975 and 1983 in line with many other species typical of farmland, but the population has remained relatively stable since then, although there was a period of decline in the 1990s. There was some range contraction in the 1970s and 1980s, especially from northern and western areas of Britain. Figure 2.8 shows the distribution of reed buntings in 1988–91.

2.8.6 Causes of decline

The decline between 1975 and 1983 occurred on both farmland and wetland habitats. During this period of decline, breeding success (young per nesting attempt) increased, suggesting that survival was the key factor. There is evidence that first year mortality was high during the period of decline with a weak correlation with winter temperatures (mortality higher in colder winters). The findings are consistent with a decline linked to low winter seed food availability.

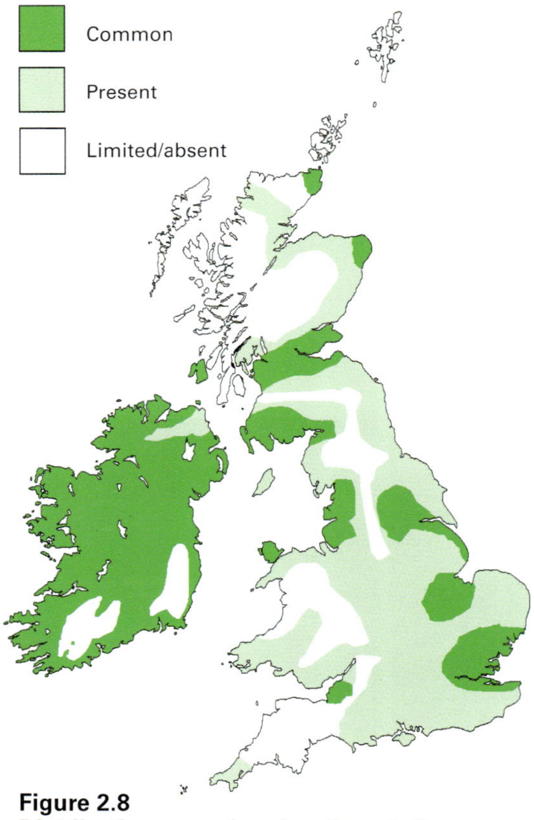

Figure 2.8
Distribution map showing the relative abundance of the reed bunting in Britain and Ireland in 1988–91.

Loss and degradation of breeding habitat, through land drainage and vegetation clearance on banks of ditches and watercourses, may also have influenced numbers. A decline in the availability of wet features on farmland may have influenced a number of species on dry arable farmland, including the reed bunting.

There was a period of decline in the 1990s, which occurred when winter survival rates were relatively high, and may have been due to poor breeding productivity. Two factors may influence breeding productivity: the availability of rank vegetation to conceal nests from predators and the abundance of invertebrate prey for chicks. Nesting success is much lower on dry farmland than in wetland habitats due to higher predation rates. Nest sites in arable crops and set-aside provide less cover than nest sites in rank and emergent vegetation associated with wetland habitats. The numbers and condition of the chicks raised and fledged was related to the abundance of invertebrate food for the chicks within 100 m of the nest in a Nottinghamshire study.

2.8.7 Management advice

The key factors are the presence of thick vegetation (such as rough grassland or crops such as oilseed rape) to provide both insect foraging and nesting habitat, and the presence of seed-rich habitats for winter foraging.

Wet features

Wet features, such as ditches, watercourses and water bodies (see section 3.4), should be maintained but allowed to develop rank and emergent vegetation as ideal nesting and summer foraging habitat. The banks should be cleared on as long a rotation as possible, so that the maximum area of rank vegetation is available each summer. Cutting and dredging should occur between September and March.

Arable

Arable field margins, field corners (see section 3.9) and beetle banks (see section 3.10) could be managed to maintain thick, tussocky swards, particularly useful for insect foraging. Oilseed rape crops in the rotation appear to boost breeding densities due to the structure of the crop providing good nest concealment as well as abundant insect populations for the chicks. It is believed that desiccation by spraying rather than swathing of oilseed rape crops benefits late broods by extending the nesting season by a few weeks to allow chicks to fledge. However, in order to prevent damage to adjacent habitats such as field margins and hedges, special care should be taken when spraying a broad-spectrum herbicide with a high boom in mid-summer. Use of wide buffer strips is the best form of mitigation.

Reed buntings will forage for insects in arable crops, and so measures such as conservation headlands, low-input cereals (see sections 3.12 and 3.13) and undersown crops will boost food availability in cereal-dominated systems, especially if oilseed rape is not in the rotation. Alternatively, plots of pollen and nectar mixtures (see 3.8) may be used to boost insect abundance.

Winter foraging habitat can be provided by seed-rich overwintered stubbles (such as rotational set-aside) (see section 3.14) or wild bird covers (see section 3.7).

Grassland

The limiting factor in grass-dominated systems is likely to be overwinter seed food sources. One option would be the maintenance or incorporation of arable fodder crops (see section 3.18) into these systems to provide arable seed sources. Leaving a ryegrass ley to seed after two

Table 2.8 Effects of different forms of land management on the reed bunting.

	Nesting	Insect food	Seed food
Bank vegetation along ditches and watercourses	✔	✔	
Improved grass field margins	✔	✔	
Arable field margins and corners	✔	✔	
Bank vegetation around ponds	✔	✔	
Oilseed rape	✔	?	?
Enclosed rough grazing	?	✔	
Beetle banks	?	?	
Conservation headlands		?	
Low-input spring cereals		?	
Undersown spring cereal		?	
Pollen and nectar mixtures		?	
Rough grazing on unimproved grassland		?	?
Wild bird seed mixture		?	✔
Overwintered stubble			✔*
Rotational set-aside			✔
Arable pockets in pastoral areas			✔

✔ indicates positive benefits
? indicates potential benefits dependent on management and outcome
* especially if preceded by low-input cereal

silage cuts (ie not taking a third cut or grazing the aftermath) was found to attract large numbers of yellowhammers to feed on the grass seed, and this may also work for reed buntings in areas where these are found.

Rough grassland areas can provide the necessary nesting and summer foraging habitat, such as uncut and ungrazed field margins around silage fields, or rough vegetation on the banks of ditches, watercourses and water bodies (see see section 3.4 and also section 3.24 on field margins of grass fields). Extensive grazing on unimproved and rough grazing land may also be used for nesting (see section 3.19).

Table 2.8 summarises the beneficial options.

Further reading

Brickle N W and Peach W J (2004) The breeding ecology of Reed Buntings *Emberiza schoeniclus* in farmland and wetland habitats in lowland England. *Ibis* 146 (suppl. 2): 69–78.

Peach W J, Siriwardena G M and Gregory R D (1999) Long-term changes in the abundance and demography of British reed buntings *Emberiza schoeniclus*. *Journal of Applied Ecology* 36: 798–811.

2.9 Corn bunting

2.9.1 Habitat use in summer

Mark Hamblin/rspb-images.com

Plate 2.12 Corn bunting

The corn bunting is typically a bird of open, lowland habitats such as farmland and coastal marshes. The preferred habitats for nesting in England are cereals, set-aside with a closed, grassy cover and ungrazed unimproved grassland (now rare in the UK), whilst improved pasture is generally avoided. However, in Scotland they often nest in improved silage fields and nests are vulnerable to destruction during silage cutting. The interesting distribution that includes mixed farmland on chalk downland, as well as intensive arable land in the fens, may relate to the shared characteristic of open farmland without tall field boundaries, or may be due to the retention of spring cropping on the lighter soils, and the insect-rich ditch banks of the fenland.

Within these habitats, breeding birds are found where there is an abundance of chick-food invertebrates within 200 m of the nest site. In the South Downs study, crops used by foraging corn buntings received significantly fewer insecticides, fungicides and herbicides than crops not used for foraging, and foraging sites also held significantly more sawfly larvae, grasshoppers and harvestmen than other areas around the nest sites. Habitats selected for foraging included spring cereals, unimproved grassland and set-aside. Winter cereals and improved grassland were generally avoided. There is some evidence that winter barley is chosen for foraging in Scotland.

Nest sites tended to be found in areas with more rough grassland in the form of grass field margins around arable fields. Nests are generally not found close to tall hedgerows or woodland. Large-scale, treeless farmed landscapes are typical of areas with high corn bunting densities. This may reflect predator evasion or competition with species such as yellowhammer. High densities of corn buntings are found where arable farmland abuts unimproved or semi-natural grassland, such as on Salisbury Plain in Wiltshire. Use of set-aside is highly variable and depends on the nature of the set-aside vegetation and the abundance of invertebrates. Non-rotational set-aside can provide nesting habitat but there is a high risk of nest destruction during cutting from mid-July. Hay meadows attract corn buntings, but nesting success will be low unless the crop is left until August.

2.9.2 Habitat use in winter

Wintering flocks are nomadic, resulting in their absence from some breeding localities. There is a complete moult after the breeding season, when birds require cover, such as post-harvest stubbles, as they are only capable of weak flight at this time. Flock sizes peak in December but flocks may still be present in late April/early May. Winter flocks may associate with finches, sparrows and other buntings, and move to sources of available grain in open locations. Traditionally, birds moved to newly available sources of seed throughout the winter. For example, in the South Downs study, the most strongly selected habitats were stubble fields in early winter, cattle feed sites in mid-winter and newly sown spring cereals in late winter. In harsh winters, or when food availability is low, corn buntings will forage on spilt grain in farmyards and grain stores, though they are less inclined to do so than yellowhammers. They will also use wild bird cover crops if there is a significant grain content.

2.9.3 Diet

Adults feed on seeds, especially cereal grains. Other plant material and invertebrates form a small fraction of the diet, and the latter mainly in the breeding season. The main plant seeds taken on arable land are cereal grains, polygonums (eg knotgrass), docks and rape. Most winter faecal samples in the South Downs study were 100% cereal grains. They take grain left on the surface after harvest and during sowing. Their attraction to spring-sown cereals at a time when alternative seed sources are scarce creates a potential risk of eating treated seed, the effects of which are not fully understood.

The young are fed on invertebrates and their larvae until they fledge. Seeds (mostly unripe cereal grains) are given to chicks when invertebrate availability is low. The main components of the invertebrate diet are grasshoppers, ground beetles, harvestmen and spiders, and the caterpillars of butterflies, moths and sawflies.

2.9.4 Breeding

Males may start defending territories from November to January, even though they are late nesters and females do not start settling in the territories until between April and June. The territory size is variable, averaging from 1 to 3 ha. The onset of breeding is related to adult food availability prior to egg-laying in some areas: often the appearance of unripe grain, but may be brought forward by supplementary feeding. Birds nesting in winter cereals start up to one month earlier than in spring cereals due to the earlier appearance of unripe grain. The end of the breeding season is related to harvest time and the availability of uncut nesting habitat. Corn buntings nest on the ground in a tussock within grassland or a dense, weedy clump in a cereal field. They rarely nest up to 1.5 m up in scrub or dense bramble.

The eggs are laid between mid-May and August. In the South Downs study, most were laid between early June and mid-July, with most clutches started after this being repeat clutches following a failed attempt. Second broods were rare in this study, but are more common in north-east Scotland.

The clutch is typically of four to five eggs but may vary from three to six. The complete breeding cycle takes about 4 weeks: egg-laying – four to five days, incubation – 12 days, and hatching to fledging – nine to 13 days. After fledging, the chicks are fed by the female for a further seven to 10 days.

In the corn bunting study on the South Downs, only 25% of nesting attempts were successful. Breeding success, in terms of nest survival and chick weight, was correlated to the abundance of caterpillars, sawfly larvae, grasshoppers and harvestmen around nests. Low food abundance also increased the foraging time required by the parents.

Second breeding attempts are rarely recorded in England, even after a failed attempt. More second broods are recorded in Scottish studies, possibly because of more uncut nesting habitat or later harvesting dates, but it is not known how many are successful. Second broods are more likely in areas with winter cereals or uncut grass available for first broods and spring cereals available for second broods. They usually start 8–14 days after fledging of the first broods.

2.9.5 Population trend and distribution

The corn bunting is generally distributed in the arable-dominated areas of the UK, although the greatest concentrations appear to be more specifically found on some of the lighter soils, such as the chalk in southern England, but then again on very heavy, wet soils such as the fens, whilst some arable areas can have very low densities. The range contraction since 1970 has largely been from Scotland, northern England and south-west England. The bird is extinct in Northern Ireland and

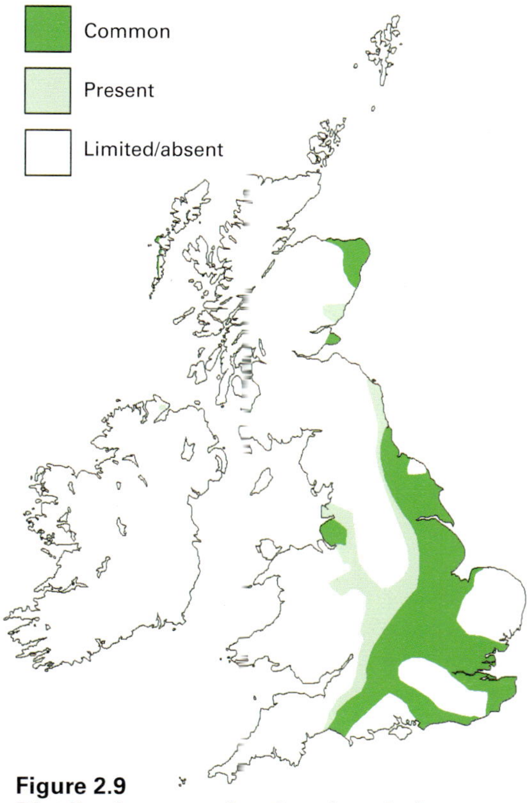

Figure 2.9
Distribution map showing the relative abundance of the corn bunting in Britain and Ireland in 1988–91.

virtually extinct in Wales. The corn bunting is believed to have declined historically before Common Bird Census surveys began in the 1960s. Survey data show that there was a large decline between the mid-1970s and the mid-1980s, and a shallower decline since then (Figure 2.9). The UK population declined by 89% between 1970 and 2001.

2.9.6 Causes of decline

Historically, the number of corn buntings in the UK was related to the area of arable, but the recent decline occurred despite the fact that the arable area increased. The species became virtually extinct in Northern Ireland and Wales during this period. The theories for the decline are:

- a reduction of winter feeding habitat due to declines in winter stubbles, outdoor fed cattle and spring cereals
- a reduction in seed availability due to the increased efficiency of harvesting and use of herbicides
- a decline in abundance of chick-food invertebrates due to an increase in pesticide use, the loss of field margin habitats and the declining use of undersown crops, which provided an abundance of sawfly larvae
- the destruction of nests during harvesting and cutting, and their influence on the number of breeding attempts possible, because the harvesting date is generally earlier now than in the 1930s
- the decline in area of late-nesting habitat such as extensive unimproved grassland.

The extinction of corn buntings from areas in the west of the UK is probably linked to the loss of arable farming. A similar loss of range resulted from the reduction of arable farming in the agricultural depression of 1870–1930. European declines have been linked to increased chick starvation and changes in cropping patterns such as the switch from cereals to maize in the Netherlands and Belgium.

In a study of the species on the South Downs, the decline was shown to be driven by an increase in pesticide use, causing poor chick survival due to low insect food abundance. Nestling weight and survival rate was correlated to invertebrate food availability, but the main direct cause of losses was predation (74% of nest failures), suggesting that increased begging due to food stress attracted predators. There was also a cost to the adults in terms of increased distance and time of foraging trips. Crop harvesting and cutting of grass and set-aside was responsible for 22% of nest failures. The proportion of nests lost to agricultural operations has increased since 1970, as the switch from spring to autumn cereals has brought forward harvesting dates, and grass improvements have led to earlier cutting.

However, British Trust for Ornithology data from across the UK has shown that breeding productivity has increased during the period of decline, suggesting that survival is the predominant factor and therefore indicating that factors affecting winter food resources are more responsible for the national decline.

In general, it is likely that both summer and winter food availability has declined with modern efficient farming systems and, whilst one factor may predominate in some areas, provision of both seed and insect food within arable farmland is required to ensure the recovery of this species.

2.9.7 Management advice

Arable

Winter seed sources are important for overwinter survival, and can be provided through retaining overwinter stubbles (see section 3.14), opting for natural regeneration on rotational set-aside (see section 3.15), or the creation of wild bird cover (see section 3.7). For corn buntings, include a cereal component in the wild bird cover mix. In one study, providing grain in the spring brought the onset of breeding forward, suggesting that food may be limited at this time. Leaving rotational set-aside unmanaged until July will provide seeds and invertebrates.

Invertebrate-rich feeding habitat in the spring is essential for chick survival. This can be encouraged on arable land through minimising the use of insecticides on crops, particularly after 15 March when many of the insects which overwinter in the field margins and hedgerows move out into the crops. Undersown cereals in the rotation provide an abundant source of sawfly larvae. Grass cover under the cereal canopy may also provide suitable nesting cover. Conservation headlands (see section 3.12) (limited use of insecticides and broad-leaved herbicides within cereal field margins) and low-input cereals (see section 3.13) are most easily adopted on light soils where the resulting weed burden will not significantly compete with the crop. Specific herbicides to tackle cleavers and grass weeds can be used with minimum impact on the food plants of the key insect groups. Alternatively, pollen and nectar mixtures (see section 3.9) can be established to create insect-rich foraging areas.

Maximising the area of tussocky grassland around arable field margins will increase the invertebrate-rich foraging area on arable farmland, and these margins and ditch banks may also be used for nesting. Wide grass margins, especially adjacent to open field boundaries or short hedgerows will provide the best foraging habitat for corn buntings (see section 3. 9). Beetle banks (see section 3.10) are a good option for increasing the foraging area in arable fields larger than 16 ha.

Creation of native grassland in arable-dominated landscapes may boost the insect food resource. Diverse native grassland can be established on non-rotational set-aside (see section 3.15) to attract breeding corn buntings. Within the set-aside rules, cutting should occur between 15 July and 15 August. For corn buntings, it is important that this occurs as late as possible to maximise the time in which breeding can be completed. Also, up to 25% of the area can be left uncut for up to three years: this can be rotated around the set-aside area. Rotational set-aside provides important winter foraging habitat but is usually not suitable for breeding as the natural regeneration rarely achieves a dense enough grassy sward, and is generally managed before August, thus destroying most breeding attempts.

Spring cereals may be preferable to winter cereals as foraging sites. However, winter cereals produce ears earlier, which encourage earlier breeding and may equally be used for nesting if there is sufficient grass margin habitat or the crop supports sufficient numbers of invertebrates. A mixed cropping system will provide more breeding and foraging opportunities.

Grassland

Within the UK, the corn bunting distribution appears to be essentially linked to arable farming, and arable pockets are vital for their continued survival in pastoral areas (see section 3.18). Thriving corn bunting populations are often found in remnant areas of unimproved or semi-natural grasslands within mixed farming areas (see section 3.19). Conservation of these areas through agri-environment scheme funding is important. Conversely, breeding corn buntings showed no association with agriculturally improved grassland in England, although this appears not to be

Table 2.9 Effects of different forms of land management on the corn bunting.

	Nesting	Insect food	Seed food
Arable pockets in pastoral areas	✔	?	✔
Wild bird cover	?	?	✔
Rotational set-aside	?	?	✔
Unimproved grassland	?	✔	?
Arable field margins and corners	✔	✔	
Beetle banks	✔	✔	
Low-input spring cereals	✔	✔	?
Undersown spring cereal	✔	✔	
Mix of spring and autumn cereals	✔	?	✔
Ditches	?	✔	
Conservation headlands		✔	?
Arable reversion to grassland	?	✔	
Non-rotational set-aside	?	✔	
Pollen and nectar mixtures		?	
Summer insecticides		✘	
Overwintered stubble			✔*

✔ indicates positive benefits
? indicates potential benefits dependent on management and outcome
* especially if preceded by low-input cereal management
✘ indicates negative impacts

the case in Scotland. The South Downs study showed the importance of grain in outdoor feed troughs as winter foraging sites for corn buntings in mid-winter where winter grazing still occurred. Traditionally managed hay meadows may be used by corn buntings for nesting and feeding, but are not safe nest sites as cutting normally occurs before breeding is completed.

Table 2.9 summarises the beneficial options.

Further reading

Brickle N W and Harper D G C (1999) Diet of nestling Corn Buntings *Miliaria calandra* in southern England examined by compositional analysis of faeces. *Bird Study* 46: 319–329.

Brickle N W and Harper D G C (2000) Habitat use by Corn Buntings *Miliaria calandra* in winter and summer. Pp 156–164 in N J Aebischer, A D Evans, P V Grice and J A Vickery (eds) *Ecology and Conservation of Lowland Farmland Birds*. British Ornithologists' Union, Tring.

Brickle N W, Harper D G C, Aebischer N J and Cockayne, S H (2000). Effects of agricultural intensification on the breeding success of corn buntings *Miliaria calandra*. *Journal of Applied Ecology* 37: 742–755.

Crick H Q P, Dudley C and Evans A D (1994) Causes of failure among buntings in the UK. *Bird Study* 41: 88–94.

Donald P F and Aebischer N J (1997) *The Ecology and Conservation of Corn Buntings* Miliaria calandra. Joint Nature Conservation Committee, Peterborough.

Donald P F and Evans A D (1994) Habitat selection by Corn Buntings *Miliaria calandra* in winter. *Bird Study* 41: 199–210.

Donald P F and Evans A D (1995) Habitat selection and population size of Corn Buntings *Miliaria calandra* breeding in Britain in 1993. *Bird Study* 42: 190–204.

Donald P F and Forrest C (1995) The effects of agricultural change on population size of Corn Buntings *Miliaria calandra* on individual farms. *Bird Study* 42: 205–215.

Donald P F, Hines P, Jackson P, Dustow J, Hepburn I, Albon G F and Jervis S (1996) Numbers, distribution and habitat associations of corn buntings on the Outer Hebrides and Tiree in 1995. *Scottish Birds* 18: 170–181.

Donald P F, Wilson J D and Shepherd M (1994) The decline of the Corn Bunting. *British Birds* 87: 106–132.

Harper D G C (1995) Studies of West Palaearctic Birds 194: Corn Bunting. *British Birds* 88: 401–422.

Hartley I R and Quicke D L J (1994) The diet of nestling Corn Buntings on North Uist – insects not grain. *Scottish Birds* 17: 169–170.

Hartley I R and Shepherd M (1994) Nesting success in relation to timing of breeding in the Corn Bunting on North Uist. *Ardea* 82(1): 173–184.

Hartley I R, Shepherd M and Thompson D B A (1995) Habitat selection and polygyny in breeding Corn Buntings *Miliaria calandra*. *Ibis* 137: 508–514.

Mason C F and Macdonald S M (2000) Corn bunting *Miliaria calandra* populations, landscape and land-use in an arable district of eastern England. *Bird Conservation International* 10: 169–186.

Ward R S and Aebischer N J (1994) *Changes in Corn Bunting Distribution on the South Downs in Relation to Agricultural Land Use and Cereal Invertebrates*. English Nature, Peterborough.

Watson A (1992) Some post-war declines in Corn Bunting numbers in north-east Scotland. *Scottish Birds* 16: 273–275.

Watson A (1992) Unripe grain, a major food for young finch-like birds. *Scottish Birds* 16: 287.

Williams T D, Reed T M and Webb A (1986) Population size, distribution and habitat use of the Corn Bunting in the Uists and Benbecula. *Scottish Birds* 14: 57–60.

3 TECHNICAL NOTES

3.1 Hedgerows

- Maintain a diversity of hedgerow types and sizes around the farm for maximum wildlife benefits.
- Avoid trimming during the peak nesting season (March to July or August) and trim on at least a two or three year rotation where possible to boost flowering and berry crops.
- Protect the hedge from over-zealous trimming, spray and fertiliser drift, heavy grazing and root damage from cultivations.
- Planting, laying or coppicing can be used to restore gappy or overgrown hedges.

3.1.1 Wildlife use

Many hedgerows have no current agricultural function but they retain a vast wildlife value. Over 600 plants, 1,500 insects, 65 birds and 20 mammals have been recorded living or feeding in hedgerows in the UK. The plant community is largely that of open woodland, both in terms of the woody plants that form the hedge itself and the herbaceous plants of the hedge base. The importance of hedgerows for wildlife is particularly high in landscapes without woodland to provide nesting cover and feeding habitat. However, they may have an important role as wildlife corridors between woodland blocks for less mobile species in wooded landscapes.

Older hedges are generally better for wildlife because they contain a greater diversity of plant life and age structure, and the associated wildlife has had longer to develop. Some hedges date back to medieval times when they marked the boundaries of parishes or estates. These ancient hedgerows are particularly important for landscape conservation, as well as wildlife. Many of the more recent hedges date back to the Enclosures Acts of 1720–1840, or earlier non-parliamentary enclosures of the 17th century. These generally take the form of a ditch with a hedge planted on the bank made from the excavated material. These hedgerows are likely to hold more wildlife than recent plantings.

Nesting, wintering and feeding habitat for birds, mammals and insects

Hedgerows are the main nesting habitat on farmland for a wide range of farmland birds. In addition, hedgerow trees can support a further range of species, such as birds of prey, owls, crows and hole-nesting passerines such as spotted flycatchers, starlings and tree sparrows. Grass margins abutting hedges can support nesting game birds, and provide food for many species that nest in the hedge itself. Outside of the nesting season, many of these species, plus others such as wintering thrushes, will continue to use hedgerows for roosting.

Hedgerows are also a source of food for birds in terms of the berries and seeds of the hedgerow plants and the insects and other invertebrates that they harbour.

Most mammals that make use of hedgerows are associated with the grass margins of the hedge, although the dormouse is a woodland mammal that is dependent on mature hedgerows to move between woodland blocks. Mature hedgerows are also important foraging areas for bats such as the greater horseshoe bat that rarely strays more than 10 m from tree and hedge cover. Hedge banks, particularly south-facing banks, can provide basking and hibernating sites for reptiles.

Insects gain food and overwintering habitat from the hedge. The abundance of insects in the hedge is related to the diversity that the hedge provides in terms of food plants, the range and timing of flowering plants, and the variety of ages of the hedge plants.

Wider hedges are generally better for wildlife as they provide better cover from the weather and predation, and are generally more resistant to adjacent land management. The margins adjacent to the

R Winspear (The RSPB)

Plate 3.1
Some tall, unmanaged hedges around the farm can provide rich sources of pollen, nectar and fruit, as well as cover for a wide range of invertebrates and nesting birds, such as turtle doves and bullfinches. Hedgerows in wooded areas of farmland are the best to allow to grow up in this way – not hedges in areas used by nesting lapwings or corn buntings.

hedge can add another level of habitat diversity and the habitat transition from the hedge to a grass margin is important for many insects.

Overall, a farm with a wide range of hedge types will support a wider range of hedgerow wildlife. Tall, thick hedges (more than 4 m tall and 4 m wide; Plate 3.1) support nesting birds such as turtle doves, bullfinches and song thrushes. Shorter hedges (up to 2 m tall) are ideal for yellowhammers, whitethroats and grey partridges as the thick vegetation of the hedge base is important for these, and this is generally shaded out in taller hedges. Hedgerow trees, especially mature trees with nesting holes, support another range of nesting birds, but also act as perches for a range of predators and so may deter other species. The shape of the hedge is generally less important, although a thick base is important for nesting birds. If a hedge becomes leggy at the base then it becomes more accessible to a wider range of predators and many hedgerow species will avoid it.

Hedgerows as wildlife corridors

There is little scientific evidence for the importance of hedgerows as wildlife corridors, with a few exceptions (eg mature hedgerows are essential for the dormouse to migrate between woodland blocks). However, it is certain that hedgerows allow more regular movement of mobile species, such as birds and mammals, around farmland than would otherwise be the case. It is possible that they also enable less mobile species such as reptiles, amphibians, invertebrates and even plants to disperse along field boundaries and colonise new habitats. The degree to which hedgerows act as useful corridors will depend on how closely they mimic the habitats that they connect. Tall, mature hedgerows are better corridors connecting woodland blocks, hedgerows with wide grass margins are better for connecting grassland habitats, and hedges with wet ditches are better corridors for connecting water bodies or wetland habitats.

3.1.2 Practical management

Developing a hedgerow management plan is a useful exercise to assess the importance and condition of the hedges and to develop the best plan for maintenance, restoration and creation of hedges on the farm. This information then needs to be imparted to all those responsible for their management. Useful steps are as follows:

- Review the historical record of field boundaries on the farm (eg using first series OS maps) to identify particularly old hedges and consider re-establishing hedgerows along lost field boundaries. The age of hedges can also be estimated by counting the number of woody species per 30 metres (the age of the hedge is roughly estimated as 100 years old per species present). Old hedges carved out of the original woodland cover can often be identified by the flora of the hedge base, which retains some of the ancient woodland plants such as wood anemone, primrose and dog's mercury. Most of the more recently planted hedges are straight and dominated by one shrub species (except very recent plantings where a wider variety of plants are used).
- Identify the shrub species in the hedge – some species are fast growing (eg ash, willow and sycamore) and may need

annual trimming, whereas others, particularly hawthorn and blackthorn, can normally be cut on a two or three-year rotation. The species present on the farm are also generally the best ones to use when planting new hedges or gapping up.

- Map hedgerow trees on the farm – the long-term plan should be to maintain the number of hedgerow trees by planting or recruiting new trees from the hedge for every one lost, or to increase the number if the farm hedges hold fewer than is typical of the surrounding landscape character.
- Map gaps in hedges that need planting up to maintain the hedge.
- Map hedges that have become overgrown and leggy at the base and are in need of restoration management through laying or coppicing (note: tall, wide, unmanaged hedges and deadwood within hedges can be of very great wildlife value and should only be brought into management if the hedge is likely to deteriorate otherwise).
- Map any hedges that need to be fenced to protect them from overgrazing by livestock.

Protection

Hedgerows lose a lot of their wildlife value if they become gappy (plants lost and not replaced) or leggy at the base so that the thick cover for nesting birds is lacking. The lifespan of the shrubs in a hedge is dependent on their management and protection. They are prone to over-zealous trimming, spray drift of herbicides, heavy grazing, and damage to the roots by cultivation of adjoining land. Pesticide and fertiliser drift will also damage the flora and invertebrate life of the hedge bank, and is more likely to result in the build-up of aggressive grass weeds, cleavers and nettles.

Broad-spectrum herbicides used when the hedge is in leaf suppress the growth and can kill young hedgerow plants. Older hedgerow plants can be killed if the damage occurs regularly. Operations such as the desiccation of oilseed rape using glyphosate and a high boom are particularly likely to cause spray drift, and wide, unsprayed buffer strips are the best protection.

Heavy grazing will eventually kill a hedge. Early symptoms are leggy hedges with an obvious browse line. Fencing is needed to protect hedges in fields with high stocking rates. The fence should be far enough away from the hedge to allow access for the flail to trim the hedge when required. Alternative shelter should be provided if stock damage the hedge plants and field margin vegetation by using it for cover.

Root damage can be reduced by not cultivating within 2 m of the hedge base. Wider margins are beneficial to a wide range of farmland birds, mammals and insects, and to protect the more extensive roots of hedgerow trees.

Trimming

Without management, most hedgerow shrubs will grow into trees, with no branches providing cover at ground level for nesting birds and other wildlife. Trimming is the easiest way to maintain the thick structure of a hedge required by nesting birds.

The practice of trimming the hedge annually dramatically reduces the flower and berry production of the hedgerow shrubs. Flower buds form on the previous season's growth, so annual trimming will remove the vast majority of the flowering sites and reduce berry production (Plate 3.2). Even cutting every second year will remove the

R Winspear (The RSPB)

Plate 3.2
This gappy hedge was trimmed in the previous winter. Note that the flowers only form on the uncut gap faces of the hedge. Trimming every second or third year enables a much greater flower and berry crop.

R Winspear (The RSPB)

Plate 3.3 Flail cutting third-year growth on an ash hedge. Most shrubs will not grow this much in three years.

berry crop before it can be used by wildlife if it is cut in the autumn, so late winter trimming (January or February) on a 2-year rotation or autumn/winter trimming on a 3-year rotation is ideal (Plate 3.3). This is generally fine for hedges dominated by slow-growing species such as hawthorn and blackthorn, but hedges dominated by faster growing species such as ash, willow or sycamore may need annual trimming. Hedges should not be left so long that trimming cannot be undertaken with the equipment available, although many hedges would benefit from being allowed to widen, so the trimmer does not need to cut back to the line of previous cuts. Some roadside hedges require annual trimming for road safety reasons.

For hedges that can be left for two or three years between cuts, it is important that cutting is done on a rotation, so that there are uncut hedges available every year for flower and berry production, to give good nesting cover, and for invertebrates like the brown hairstreak butterfly to overwinter. Avoid leaving all the hedges on the farm for three years and then cutting them all in the same winter.

Allowing the hedge to increase in height and width slightly at each cut will create thicker nesting cover as side branches form directly behind the cut stems. The long-term consequence of this may be that more radical management is needed at some stage, but if the hedge is allowed to expand by a few centimetres per cut then it would take 50 cuts (over 100 years if using a two to three-year rotation) to expand by a metre. If this is not practical across the farm as a whole, look for situations where it is more convenient, such as field corners or hedges with wide verges. Also, consider the needs of species such as yellowhammers, which use hedgerows of less than 2 m. They may be accommodated by maintaining the height, but allowing the hedge to widen out.

Trimming should not be done during the peak nesting season, which is largely from March to July inclusively, although many species continue through August too (see Appendix 1). Trimming in late winter, to allow wildlife to make use of the berry crop, is less important if trimming is done on at least a 3-year rotation, so that at least two-thirds of the hedges are left uncut each winter.

When thick growth is cut with a flail trimmer, the resulting stems look badly lacerated, but most hedge plants can recover from this. However, a neater cut is less likely to damage the plants. This can often be achieved by keeping the flail in good working order and taking care when deciding how far back to cut. Cutting thicker growth and branches of hedgerow trees is best done with machines using a circular saw rather than a flail.

Hedge shape is not as important for wildlife as width (the wider, the better) and height (a variety in keeping with the landscape character is ideal). Allowing hedgerows to increase in height is often more beneficial in wooded landscapes where they may act as more effective corridors for wildlife. It may be disadvantageous to increase the height of hedges or plant hedgerow trees in open farmland areas used by nesting birds such as lapwings and corn buntings, which avoid nesting near tall field boundaries.

Laying and coppicing

Many hedges that have been allowed to grow without trimming become leggy and cease to provide cover at ground level. Laying or coppicing can restore this type of hedge. These are labour-intensive operations, but were only undertaken approximately every 20 years when they were the dominant forms of management, and the length of the cycle can be significantly increased if done in conjunction with trimming management.

Hedge-laying (or steeping) is possible for relatively young gappy or leggy hedges, but if the hedge consists of very thick stems then coppicing is a better method of restoration. The ideal diameter of stems for laying is 10 cm. If large gaps have been created, then coppicing and planting up the gaps is the best alternative. Laying and coppicing should be done when the plants are dormant (not in leaf). Some stems can be left to grow up into hedgerow trees, and it is better to leave some standing deadwood for the benefit of invertebrates.

The basic principle of hedge-laying is to cut through the stems (pleachers) until they can be bent over almost horizontal (but not quite, as sap flow still needs to rise) along the hedge-line. The hedge should be laid uphill if it is on a slope. The resulting hedge is often more open in the first year, but thickens up rapidly in subsequent years (Plate 3.4). Prior to laying, a hedge should be allowed to grow up sufficiently to provide enough length for effectively weaving the pleachers together when they are laid. Much of the wood is removed from the hedge during laying, otherwise the bulk of the hedge would be too much to allow the pleachers to be lowered. Species that

R Winspear (The RSPB)

Plate 3.4 Hedge-laying is a good way of restoring thick, bushy growth at the base of the hedge.

do not respond well to laying, such as elder, are generally removed, as well as many of the side branches of the pleachers. If removed wood can be stacked somewhere that does not interfere with farming operations then it creates a very valuable wildlife habitat. Further practical detail can be found in Brooks and Agate (1998; see Further reading below).

Coppicing is the best means of restoration if the hedge is too gappy, or the stems are too thick to be laid. It involves cutting the hedge plants down at just above ground level (the lower, the better and no more than 10 cm above ground level). The cuts should be angled so that rainwater runs away from the centre of the coppice 'stool' to reduce the risk of rotting. Gaps can be planted with new shrubs after coppicing so that they are not in the shade of the existing plants.

Trimming the new growth of laid and coppiced hedges on a two or three-year rotation creates more side shoots and thicker growth. On laid hedges, cutting should follow the direction of laying. Fencing may be necessary to protect the hedge from grazing livestock. The new growth may also be more vulnerable to herbicide spray drift. The regrowth will be more vigorous than new plantings, so weed control should not be necessary unless gaps have been planted up.

Coppicing and laying will reduce nesting opportunities for a few years after management and should therefore be done over many years, rather than managing large sections in one year.

Planting and gapping up

New hedges are best placed along old boundary lines or to link existing habitats such as woodlands or areas of scrub. Even an old hedge-line with a few remnant hedge plants may be worth restoring. The species found in existing hedges are the best to use, as these are more likely to thrive, and the associated invertebrate life will be available to colonise the new hedge. Local plant sources should be used wherever possible to conserve the local

R Winspear (The RSPB)

Plate 3.5
Hedge planted through polythene to retain soil moisture and control weeds around the plants.

genetic stock. This may also provide plants better suited to the conditions. The wildlife diversity of the hedge will be related to the diversity of shrub species planted. Aim to use at least five shrub species typical of hedgerows in the area.

Where possible, it is best to plant on a raised bank (at least 0.3 m high and 1 m wide) to provide free-draining habitat for insects and ground-nesting birds. However, planting on a bank increases the risk of drought killing the hedge, so do not create a bank in situations where soil moisture is likely to be limited. Planting should be done between October and March, but not when the ground is frozen. Autumn is the best time, when the soil is still warm and there is more moisture available. Plant up to eight plants per metre using a staggered double row with about 25 cm between plants in the row and 20–30 cm between the rows. It is generally recommended to plant groups of the same species together to reduce the risk of less competitive species being out-competed by more competitive neighbours. Weed control is necessary before planting and for at least 3 years afterwards, to allow the shrubs to establish successfully. Non-chemical options include using fertiliser bags or black polythene (Plate 3.5), which can retain moisture for the hedge as well as protecting it from weed competition. Straw is a natural alternative. The plants usually need protection from grazing using rabbit guards or fencing off if livestock are present. Trimming the top during the winter after planting, and then annually until it reaches the desired height, will encourage more bushy growth.

Gapping up existing hedges is best done in conjunction with coppicing to reduce competition for the new plants from existing plants in the hedge.

Further reading

Barr C, Britt C, Sparks T and Churchward J (2005) *Hedgerow Management and Wildlife. A review of research on the effects of hedgerow management and adjacent land on biodiversity*. Department for Environment, Food and Rural Affairs, London.

Brooks A and Agate E (1998) *Hedging: a practical handbook*. British Trust for Conservation Volunteers, Doncaster.

FWAG (1999) *The Good Hedge Guide*. Bayer plc Agricultural Division.

Sotherton N and Page R (1998) *A farmer's guide to hedgerow and field margin management*. Game Conservancy Limited, Fordingbridge.

3.2 Hedgerow and in-field trees

When undertaking any work on mature or ancient trees, follow this checklist:

- Seek specialist advice to determine the best management.
- Check that the tree is not protected under a Tree Preservation Order.
- Check that all work complies with relevant wildlife law.
- Protect trees from damaging farming operations, poaching from livestock, drainage, neglect and inappropriate management.
- Wherever possible, retain standing decaying wood. If necessary, seek to reduce the canopy of dead standing trees where possible, rather than fell it completely.
- Leave any cut or fallen timber intact and *in situ* wherever possible.
- Create a population of varying aged trees to maintain a sustainable supply of veteran trees for perpetuity.
- Plant new trees and bring existing middle-aged trees into management. When planting new trees, wherever possible always plant native trees. However, consider using shorter-lived non-natives as a stopgap if tree numbers are very low.

3.2.1 Wildlife use

Hedges in the UK contain mature, over-mature and ancient trees (Plate 3.6). The Countryside Survey 2000 found that 30% of the hedge tree population was over 100 years old. Many in-field trees are remnants of old hedgerows and have the same intrinsic wildlife value. A lot of the wildlife value of farmland trees is a result of the decaying wood component. This provides habitat for a wide range of fungi, lichens and invertebrates, and roosting and nest sites for birds, bats and other small mammals. Over 1,800 species of invertebrates are dependent on decaying wood. This represents about 6% of the invertebrate fauna of the British Isles.

R Winspear (The RSPB)

Plate 3.6 The ash trees in this hedge provide nesting holes for a population of tree sparrows.

3.2.2 Practical management

There is a lot of current research looking at the management of ancient trees. It is therefore vitally important that a specialist is consulted to understand the latest management advice. A list of useful contacts and information sources are given at the end of this section. Before undertaking any management, check that a Tree Preservation Order does not protect the tree. A specialist survey of the state of the tree and its associated wildlife should also be carried out. It is important to note that under the Wildlife and Countryside Act and The Wildlife Order (in Northern Ireland), all species of bats are fully protected at all times, as are the active nests of all birds during the breeding season.

Hedgerow tree management

Old trees should be retained in hedges for their exceptional wildlife value. Standing decaying wood is also important for fungi and invertebrates and, if it has suitable holes, nesting and roosting birds and bats. It is important to leave standing dead trees and dead branches of mature trees wherever it is safe to do so. The lifespan of hedgerow trees can be reduced by root damage from cultivation or trampling from livestock and lower water tables following field drainage.

Uncultivated margins along hedges with trees, that are at least as wide as the tree canopy, provide greater protection to the roots from cultivation damage. Scattered scrub growing beneath mature hedgerow trees will also protect the roots and canopy from field operations. Scattered scrub and ruderal vegetation also protects fallen and standing decaying wood from desiccation by providing shade and humid conditions as well as an important nectar source for invertebrates. Prevent any scrub from becoming too dense as this will compete with the tree and reduce its growth rate. Too dense shading of deadwood is also damaging, as warmth from sunlight can be important to larval development. Heavy shading of tree trunks is also damaging to lichen interests.

Pollarding was a traditional method of harvesting wood and leaf fodder from trees whilst stock grazed beneath the pollards without grazing the regrowth. The distribution of pollards across the UK is still not fully known. However, they are well known features in the Lake District, Cotswolds and parts of Hertfordshire. Old pollards can be identified from the distinctive sprouting of branches at a particular height. The crown of the tree was cut at between 1.8 and 4.5 m high. This was repeated on a rotation of seven to 15 years under formal cutting regimes. Casual pollarding of selective boughs, mainly for roofing and ship beams, also took place.

Pollarding often allows a tree to reach a much greater age than if left to grow normally. Continued management of old pollards, and establishing a successor generation where specimens have been lost, is important. Maintaining well-lit trunks is important for lichens and helps maintain the wood-decay conditions required by the invertebrate communities of these mature trees. Reinstatement of old pollards is very difficult and many specimen trees have been lost as a result, through lack of knowledge and understanding of the risks. Such restoration should only be carried out after seeking specialist advice. Table 3.2 indicates the susceptibility of certain tree species to pollarding.

Before reinstating pollarding or creating new pollards purely for historic reasons, consider the conservation objectives and

Table 3.2 Vulnerability of species to pollarding.
Taken from Forbes, V and Clarke, A (2000) Bridging the Generation Gap. *Enact* 8 (3): 7–9. English Nature, Peterborough

Species	Vulnerability	Max dbh (diameter at breast height) at which pollarding may be carried out in one attempt
Hawthorn	1	20 cm
Field maple	2	16 cm
Hornbeam	3	16 cm
Crab apple	4	Unknown
Ash	5	10 cm
Elm	6	Unknown
Oak	7	8 cm
Beech	8	Unknown
1 = easy to pollard, 8 = difficult to pollard		

potential threats that may be brought to the tree. Trees older than 50 years, including lapsed pollards, are likely to respond better to phased cutting over several years. Varying the cutting frequency, number of poles cut and the height of the pollard are just some of the techniques that might be considered. Trees whose crowns have become unbalanced may require preventive tree surgery to retain the long-term interest. Selective thinning or reduction of boughs may be necessary.

Many of the insects associated with decaying wood also require pollen and nectar sources in their adult stage, and flowering trees, shrubs and herbaceous plants in the adjacent hedgerow and its margins can be vital.

It is important to retain a sustainable population of ancient trees. In order to achieve this, aim to manage a range of different-aged trees throughout the landscape. Old trees can be replaced in the long-term by new plantings or by selecting saplings within the hedge to grow up. The ideal saplings to select are those of tree species found elsewhere on the farm that have not been trimmed in the past. Trees grown from saplings that were trimmed, coppiced or laid will be damaged at the point where they were cut. Good saplings are best found in a hedge that has not been trimmed for two or three years, or from scrub allowed to grow beneath a mature tree. Hedgerow saplings should be clearly tagged so that they are avoided during future hedge trimming operations. The tagged saplings can be made more visible for the hedge-cutter by hand trimming the hedge 1 m either side of the sapling. Selecting a few saplings each year will spread the age structure of the trees over time.

If planting is the preferred option, then trees should either be planted in existing gaps in the hedge, included when planting a new hedge or planted after coppicing to reduce competition from the hedge. Again, native species regularly found in the area are the best to use. They vary in the range of wildlife that they support. Oaks, which are widely found in the UK, have the greatest range of associated insect life. Other species commonly found in some areas include ash, beech and holly. Elm is still an important hedge tree, but it rarely survives beyond about 15 years old, when it becomes vulnerable to the beetles that carry the Dutch elm disease fungus.

It is estimated that 50% of newly planted trees die within the first five years due to a lack of aftercare. The following checks are all important to ensure the survival of the trees:

- checking that planted trees are vertical
- removing stakes and ties once the tree roots are stable
- draining waterlogged soils around the roots
- firming the soil around the roots if it has been lifted by winter frosts
- remove vegetation around the trees if it is likely to out-compete them for light or moisture.

While planting provides for the long-term supply of mature trees, it is important to retain middle-aged trees around the farm for ecological continuity. Ideally, these should be native trees, but where this is not possible, use a suitable non-native species as a stopgap. Alternatively, enhance any planting with faster growing trees such as sweet and horse chestnut, and birch. These species have shorter life spans and will provide a decaying wood habitat more quickly.

Protection of in-field trees

In-field trees warrant the same protection as hedgerow trees. Management threats to in-field trees come from:

- felling of either the whole tree or its boughs for safety reasons, tidiness or changing land use
- neglect (usually in the case of old pollards)
- damage during unskilled tree surgery (but see below)
- ploughing, spraying pesticides or applying fertiliser underneath the canopy and compaction caused by driving of vehicles over the roots
- high stocking rates leading to bark stripping, soil compaction or nutrient enrichment
- changes in water levels leading to drought or waterlogging.

In addition, the associated wildlife can be affected by:

- the isolation of the tree from other trees and the lack of new generations of trees to allow less mobile species to spread to and from the tree
- over-zealous tree surgery, that removes decaying wood throughout the whole tree canopy – dead boughs are very important to many specialist invertebrates and lower plant communities, as well as providing an important feeding area for birds
- the drift of pesticides and fertilisers onto or below the canopy.

Herbicide use (other than spot treatment of pernicious weeds) should be avoided to protect plants such as lichens growing on the trunks. Fungicides and inorganic fertilisers can damage the mycorrhizal fungi that enable the tree to take up nutrients and protect it from disease. Insecticides can remove the associated insects. Even the residues of veterinary products voided in dung and urine have been implicated in the decline of mycorrhizal fungi, especially where antibiotics are in routine use on the livestock. Decaying wood should be retained where it is safe to do so, and all fallen timber left *in situ* where possible or moved a minimal distance toward the base of the tree so as not to interfere with farm operations. This may also help protect the trunks and roots from livestock damage.

The roots should be protected from cultivations and compaction from stock trampling or heavy machinery. As such, operations such as cultivations and supplementary feeding and watering of stock should not take place under the canopy of the trees. Ideally, the area under the canopy plus 5 m, or 15 times the diameter of the trunk at breast height, whichever is the greater, should remain uncultivated. As with hedgerow trees, ruderal vegetation and scattered scrub provides protection and prevents desiccation of the timber.

When planting for future replacement of open-grown trees, look to plant in groups of three at 12–15 m spacing to avoid branch contact at maturity.

The removal of limbs by tree surgery is generally only advisable to restore an old pollard, in the management of existing and new pollards, or to manage a tree that is liable to fall on a site with an overriding issue of public safety. When removing limbs for safety reasons, consider a reduction rather than complete removal. Jagged edges have greater wildlife value than straight cuts. Besides having a more natural appearance, they provide habitat for fungi and invertebrate communities. Where possible seek to snap rather than cut limbs. Sideways or upward winching is more likely to achieve the desired effect. Seek specialist advice before undertaking such work.

Further advice and reading

The Tree Council has a range of member organisations that can provide specialist advice on the conservation of old trees. Details can be obtained from www.treecouncil.org.uk.

Ancient Tree Guides No 1: Trees and Farming. Woodland Trust and Ancient Tree Forum. Ancient Tree Forum: http://www.woodland-trust.org.uk/ancient-tree-forum/

English Nature (2000) *Veteran Trees: a guide to good management.* English Nature, Peterborough.

Forbes V and Clarke A (2000) Bridging the generation gap. *Enact* 8 (3): 7–9.

Stokes J and Hand K (2004) *The Hedge Tree Handbook.* The Tree Council, London.

3.3 Scrub

Scrub is an important habitat for birds and other wildlife. However, scrub encroachment can also be a threat to important open grassland habitats and the landscape character. Careful management ensures that scrub does not compromise or threaten farming, wildlife, archaeological or landscape interests. The following factors should be considered:

- Scrub management should aim to create, maintain and enhance the variety of age and structure of the shrubs.
- Prevent unwanted establishment early on. Eradication of established scrub is more difficult and techniques more constrained, especially on archaeological sites.
- Consult specialist organisations where appropriate.

3.3.1 Wildlife use

Scrub is an important wildlife habitat, whether it is a few isolated shrubs or young trees, or a dense thicket. It is a natural part of open habitats, such as grassland (Plate 3.7) and heathland, and an important component of the landscape. When well managed, it is thinly spread across these open habitats and will have a variety of ages, structures and plenty of edges that, in turn, will increase the diversity of wildlife. Scrub can also help to buffer farm woods and ditches from spray drift.

Plate 3.7 Scrub developing on semi-improved chalk grassland, providing nesting opportunities for a range of birds.

R Winspear (The RSPB)

Soils and climatic conditions influence the types of scrub that develop. Hawthorn is generally the most widespread species, growing mostly on neutral, chalk and limestone soils. Blackthorn dominates on the deeper, more fertile soils; gorse, broom and bramble on acidic soils and willows dominate on damper soils. Table 3.3 gives a brief overview for some of the main scrub types likely to be found. However, because of the nature and complexity of scrub, this is not exhaustive.

Well-managed scrub and its margins support a range of wildlife. Scrub provides nectar, pollen, seeds, fruits, shelter and nest sites for invertebrates, birds and mammals. It also offers suitable habitat for many herbaceous plants.

Scrub of varied age, species and structure supports the widest range of wildlife, as some species depend on specific growth stages of certain plants. Some species require particular shrubs and others a range of habitats within a small patch of scrub. It is important to maintain all growth stages, from bare ground through young and old growth to decaying wood.

Scrub edges are an important habitat, often rich in flowering plants. These provide pollen and nectar for insects, and seeds for birds and mammals. Tall herbs and grasses growing along the edge of scrub offer shelter for small mammals, nest sites for birds and hunting areas for barn owls and kestrels.

Birds nest in a range of scrub types. Yellowhammers, linnets, grasshopper warblers and whitethroats favour young, scattered scrub. Dunnocks and willow warblers use low-growing, closed-canopy scrub. Turtle doves, song thrushes and bullfinches use older, mature stands of scrub. Nightingales require very dense stands of, for example, blackthorn or brambles. Table 3.4 gives examples of the various niches used by birds.

Wet scrub is important to a range of invertebrates, such as weevils, craneflies and moths. Its structure can provide shelter

Table 3.3 Main scrub types found in the UK.

Scrub types	Description	Location
Hawthorn	Found on most neutral and base rich soils. A mix of trees, shrubs and climbers	Widespread throughout the UK
	Species-poor – mostly hawthorn, blackthorn, elder, bramble and rose. Sometimes diverse with ash and field maple. Some regional variation in west and north where sycamore occurs and in the Midlands and East Anglia where elm suckers occur	Widely distributed throughout much of lowland England, especially on disturbed or made-up ground, landslips, abandoned arable or grazing land, and on mineral workings
	On calcareous soils it is high in species-richness, and includes among others: spindle, wayfaring tree, blackthorn, roses, privet, dogwood	Restricted to free-draining soils over limestone in the Derbyshire Dales and the Lake District, and chalk downs in south-east England
Blackthorn and bramble	Forms a closed canopy that excludes most other species, mainly on deep, moist, neutral soils. On poorer soils gorse often occurs, while hazel and privet occur on more base-rich soils.	Widespread throughout lowland UK. Develops on abandoned grassland on woodland edges. Also exposed places, such as cliffs and landslips, where it may form a stunted, almost climax community
Bramble and coarse grass	Early successional scrub; bramble usually occurs as dense clumps with the coarse grasses growing between	Occurs widely on abandoned and waste ground throughout much of the UK. Also at the edge of woodlands, as well as along hedges and transport corridors
Bramble and bracken	Deep, free-draining, neutral to slightly acid soils. Bramble dominates with scattered hawthorn, blackthorn and elder. Occasionally with sapling ash, oak, beech and sycamore	Widespread underscrub community, mostly typical of woodlands. In appropriate soil conditions, it can appear throughout lowland UK
Gorse and bramble	Typical community of acidic, free-draining soils on lowland and upland fringes. In the absence of burning or grazing, the stands can grow very tall and leggy	Widely distributed on suitable soils on marginal land, along boundaries of enclosures, along tracks and hedge banks
Willow	Found at the margins of standing or slow-moving water	Widely scattered throughout the UK
	Wet carr scrub of mires. Mainly secondary succession onto abandoned marshes. Also develops by direct invasion onto herbaceous fen	Confined mainly to the Breckland and Broads of East Anglia, and the Cheshire Meres
Alder	Wet alder carr of swamp, fen and flood plain mires on organic, base-rich soils	Local but widespread in lowland UK, mostly in north-west Britain, the Weald and parts of East Anglia
	Wet alder carr scrub on alluvial river terraces or flood plain mires. Under suitable conditions occurs on enriched soils over acid mires	Widespread but local throughout the lowlands, mostly in the north-west and East Anglia

Table 3.4 Examples of scrub features used by some birds of conservation concern.

Species	Scrub features			
	Edge	Scattered	Open canopy	Dense canopy
Bullfinch	✔			✔
Cirl bunting	✔	✔	✔	
Dunnock	✔			✔
Grasshopper warbler		✔	✔	
Linnet	✔	✔		
Nightingale	✔			✔
Song thrush	✔	✔		✔
Turtle dove	✔			✔
Willow warbler		✔	✔	
Yellowhammer	✔	✔	✔	

and warmth within more open habitats. Many species of birds benefit from wet scrub, including warblers (Cetti's, grasshopper and marsh), the reed bunting, the willow tit and, in mature areas, the lesser spotted woodpecker. Wet scrub can provide otters with lying-up sites or holts, and the high humidity favours the growth of mosses, lichens and fungi.

3.3.2 Practical management

There are many opportunities for scrub development on farmland (Figure 3.1). It can spread from hedges or woodland edges into field corners and edges that are awkward to manage. Scrub that develops along a woodland edge is valuable as it creates a rare transition between wood and open ground. Larger stands can be created in field corners where two or more hedges intersect. Scattered scrub growing beneath mature hedgerow trees or in-field trees helps protect them against damage from soil compaction over the roots. However, do not allow it to become too dense, as this will compete with the growth rate of the tree (see section 3.2 on hedgerow and in-field trees).

Left unmanaged, scrub will develop into woodland. It requires periodic maintenance to retain its character and value to wildlife. It may also need managing to prevent it from damaging farming, wildlife, archaeological or landscape interests. Long-established scrub enriches the soil and if it is removed, competitive weeds will grow. It would take many years for the area to return to

Table 3.5 Summary of techniques for managing scrub.

Technique	Objective		
	Create	Restore	Reduce
Natural regeneration	✔		
Planting	✔		
Helping plants establish	✔	✔	
Protection from browsing	✔	✔	
Grazing and browsing		✔	✔
Cutting scrub		✔	✔
Mowing and flailing		✔	✔
Stump removal and grubbing out		✔	✔
Herbicide application	✔	✔	✔

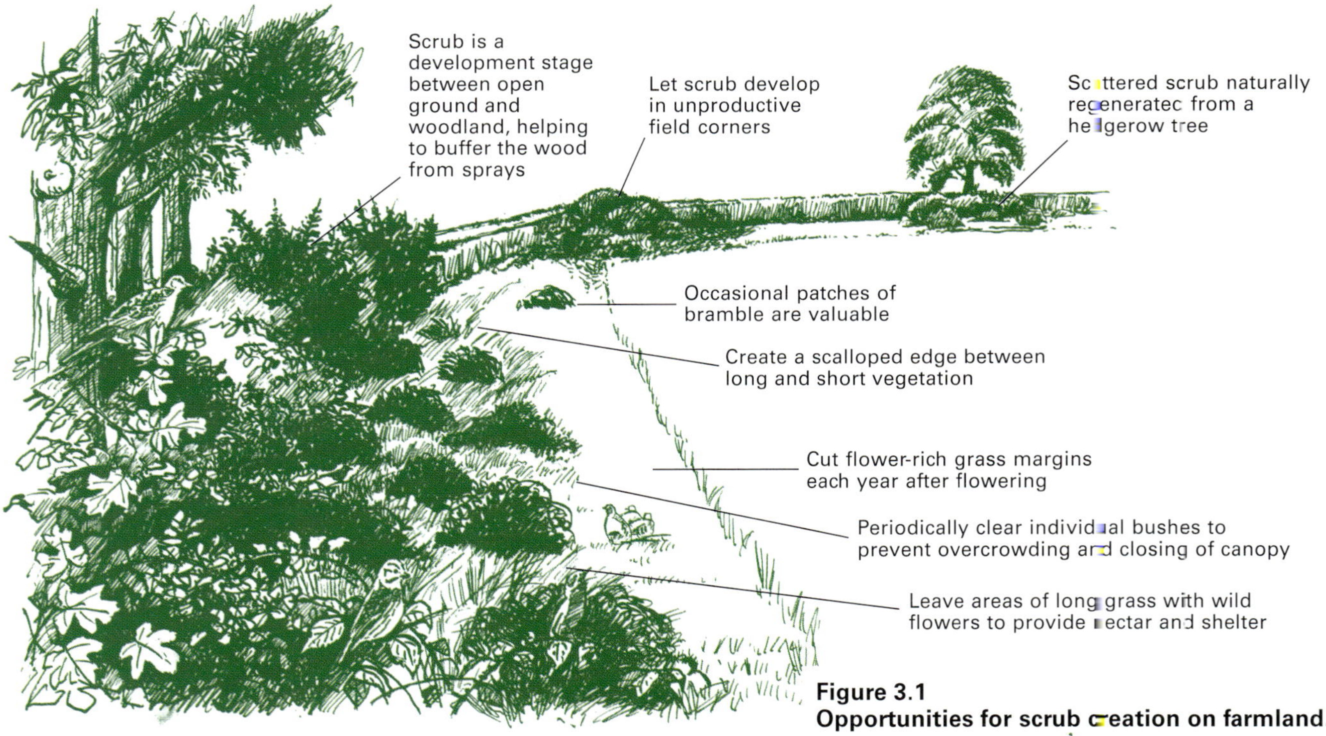

Figure 3.1
Opportunities for scrub creation on farmland.

flower-rich grassland. Consider the impact of management on the landscape, and on species-rich grasslands and archaeological sites in particular.

Scrub is a dynamic habitat that naturally colonises new areas and benefits from rotational management. To maintain the desired balance, remove some older stands, while allowing younger, regenerated growth to spread where acceptable. Encouraging scrub to move around a site in this way will retain age and structural diversity and benefit species requiring the pioneer stages of scrub development. Where it is not possible to move scrub in this way, use browsing and/or rotational cutting to maintain stands.

There is a wide range of techniques for managing scrub to meet different objectives. These are summarised in Table 3.5.

Creation or development

Natural regeneration:

- Scrub develops naturally if allowed to do so. Allow scrub to colonise where there are no threats to farming interest, wildlife, archaeological or landscape features.
- Excluding grazing or stopping cultivation is often all that is needed to encourage regeneration. The scrub will probably develop slowly, and attract a variety of wildlife through the various stages of its succession.

Planting:

- Planting should only be considered as a last resort, where there is no natural source of regeneration or where a rapid result is required. Native species should be used, and preferably plants should be of local provenance, either from locally collected seeds or 'whips' (nursery-grown saplings).
- Planting between November and March ensures the maximum survival of new plants.
- Try not to plant in rows, as this creates wind tunnels. Scallop the edges of stands; mix species randomly to create diversity or plant in clumps to create a natural appearance. Leave unplanted gaps, which will infill naturally.

Helping plants establish and protection from browsing:

- Scarification of compact ground will help species with light seeds to establish. Weeding when seedlings or whips are establishing will remove competing vegetation. Using chippings or straw to mulch about a metre around each plant will suppress weeds and help to retain soil moisture.
- Use herbicides only if there are no alternative methods of control. Spraying up to a 1 m radius around each plant will suppress weed competition. Avoid broad-spectrum contact herbicides – consider a graminoid herbicide to control grasses, but avoid wild flowers, as they are important for nectar-feeding invertebrates. It is a legal requirement to follow all product use labels and comply with all statutory regulations.
- Palatable shrubs are vulnerable to livestock, deer and rabbits during establishment. Small numbers can be protected with spiral guards or tubes. However, these are liable to cause the plants to have a lollipop appearance and poor structure. Larger areas will need fencing to protect them from rabbits and/or deer as required. In fenced stands, gated access will allow for controlled, short bursts of livestock grazing at appropriate times and density. This will help keep the grass sward open and improve the structural diversity of the scrub.

Restoration and maintenance

Aim to create or enhance the following features:

- sunny, sheltered edges, which offer a hot micro-climate that is important for insects
- scalloped edges that increase the length of edge and provide shelter
- rides through scrub that provide sheltered edges – avoid openings that face the prevailing wind
- patchworks of scrub and glades that provide a lot of edge
- bramble, which is valuable for nesting and feeding birds and for nectar-feeding insects
- dead wood, which is valuable to fungi and invertebrates – leave dead trees or shrubs standing and retain small stacks

of cut wood in partially shaded areas to rot slowly
- bare ground, which is valuable for insects and scarce plants – you can create it by uprooting scrub, or with a digger, but avoid creating it in areas of archaeological importance.

Grazing and browsing:
- Light grazing helps to maintain the scrub edge and glades by suppressing rank grasses. Browsing improves structural diversity, but can damage palatable species. Different species can be palatable at different times of year.
- Grazing and browsing behaviour differs between breeds. Most cattle do not browse, but can open paths through scrub by trampling. Sheep generally prefer succulent, young growth. Some breeds of sheep are good at browsing and pushing through scrub, but younger animals and lighter breeds are prone to becoming caught up in it. Ponies graze and browse and will strip buds and bark in winter or early spring as well as open paths. Goats can strip bark and, if used carefully, will produce structural diversity.
- Try to achieve a balance between diversifying the scrub stand and its margins, and causing damage. Start with a low stocking rate (about 0.25 livestock units (LU)/ha) and monitor the effects. Increase stock numbers if necessary to achieve the desired balance.

Cutting scrub:
- Cutting most species of scrub encourages re-growth, and is useful for maintenance and restoration. Cut areas of scrub in a rotation, aiming to retain all ages. Scrub typically matures in 15 years, so cutting one-fifteenth every year or three-fifteenths every third year, for example, will restore a good age structure.
- Cutting small patches will diversify scrub structure. Avoid cutting adjacent patches sequentially, as this reduces foliage available for invertebrates to feed on.
- Plan the cutting programme to take account of existing interests and constraints (archaeology, landscape and wildlife).
- Cut between September and February, to avoid the peak nesting season.
- Leave berry-bearing scrub until after Christmas for birds and mammals to feed on.
- Suitable tools range from hand tools – such as bow saws, mattocks, chainsaws and brush cutters – to tractor-mounted flails, depending on the size of the task.
- Avoid creating access routes and burning cut material on or near sites of wildlife or archaeological importance.
- When cutting or clearing wet scrub, it may be necessary to use marsh mats to protect sensitive vegetation and prevent sinking into the soft ground. Where possible, retain patches of open wetland, such as pools or marsh.

Edge management by mowing or flailing:
- Occasional mowing or flailing will maintain rides, glades and scrub edges.
- Annual mowing will keep the grassy scrub edges open and encourage flowering herbs.
- Avoid destroying seeding herbs by delaying mowing until late summer/ autumn.

Stump removal:
- Stumps are important for wildlife, so should be retained where possible. However, removal may be necessary during ride or glade creation to allow access for mowing.
- Do not remove stumps in areas of archaeological importance.

Herbicide application:
- Herbicides can be used to control weeds to assist establishment, as well as controlling regeneration or removing scrub.
- Herbicide can be applied to scrub re-growth by spot spraying or weed wiping. The latter is usually necessary where sensitive non-target plant species are present. Freshly cut stumps may also be painted with herbicide.
- In keeping with Government policy to minimise the use of herbicides, always consider the need against other available methods. Always use the approved herbicides and follow the on-label instructions for use.

Reduction or eradication

Scrub encroachment needs to be prevented where it is detrimental to farming, conservation, landscape or archaeological interests.

- Cutting can be used to reduce or eradicate scrub. Follow-up treatment will be required– for example, repeated cutting, mowing, continued browsing of re-growth, removal of stumps or herbicide treatment. Use herbicide application to treat stumps or re-growth where there are no alternatives.
- Use of large machinery is often necessary to tackle severe cases of scrub encroachment. Old or unwanted scrub can be removed quickly and efficiently with a digger. This removes the scrub, its roots and nutrient-rich top soil to re-expose the mineral soil and restart natural succession.
- Do not use a digger where it is likely to damage wildlife interests or archaeological sites. In these situations, scrub will need to be cut.
- Stumps are important for wildlife, especially fungi and insects. They should be left, except where access or maintenance such as mowing is required. Removing stumps from archaeological sites is likely to cause damage.
- In some circumstances, wet scrub may be killed or suppressed by raising water levels.

Preventing scrub establishment: on sensitive areas and for certain species, it is important to prevent the establishment of scrub. This particularly applies to introduced species.

- Grazing will suppress scrub establishment. The stocking rates needed to maintain the grassland is usually sufficient (about 0.25 LU/ha), but may need to be increased to tackle a particular problem.
- Mowing can be used in conjunction with grazing or where grazing is not possible. Avoid mowing until late summer/autumn to allow time for flower and grass seeds to drop, or late winter/early spring to give overwinter shelter for insects.
- Where scrub has begun to colonise, remove seedlings and their roots immediately, as heavy infestations are harder to eradicate.

Difficult to manage species: some species – in particular, blackthorn, dogwood and privet – are difficult to remove and will spread vigorously in response to cutting.

- Where necessary, and if there are no limiting factors (eg access, archaeology, landscape or wildlife), grub them out using a digger. Where this is not possible, cut the scrub and weed-wipe the re-growth with herbicides.
- Goats can control regeneration, usually within two or three seasons.

Juniper

Some lowland farms on calcareous soils may have juniper scrub. Juniper is a rare shrub of calcareous soils. Where it does survive, it is usually as old, single-aged stands. Regeneration from seed is vital to ensure the survival of the stand and diversify its age and structure.

- Juniper seeds in open, disturbed ground. Such areas adjacent to existing juniper offer ideal conditions.
- Protection of seedlings from browsing rabbits, deer and livestock is vital.
- Juniper is slow growing and can suffer from crowding and shading by other shrubs. Selective management is needed to maintain healthy juniper bushes.
- Overgrazing is the main cause of poor regeneration. However, fencing stock into small enclosures for short bursts of light grazing will create the open sward conditions that are ideal for seedling establishment. Monitor the effects and move the enclosures and stock to the next target area as required.

Where juniper occurs, seek specialist advice from Plantlife International, 14 Rollestone Street, Salisbury, Wiltshire SP1 1DX or visit www.plantlife.org.uk.

Further reading

Day J, Symes N C, Robertson P A, Bacon J (eds) (2003) *The Scrub Management Handbook: Guidance on the management of scrub on nature conservation sites.* FACT, in association with English Nature. Download free from www.english-nature.org.uk/pubs/handbooks

3.4 Ditches, watercourses and water bodies

Ditches should be cleaned on a little and often rotation and a variety of channel and bank habitats should be maintained around the farm. Watercourses and water bodies containing clear water, a range of plants and an abundance of insects are signs of good water quality. They should be protected from pesticides, fertilisers and soil erosion.

3.4.1 Wildlife use

Few farms do not contain at least some wet features such as streams, drainage channels, ponds or seasonal pools. However, wet features on farmland have become far scarcer with the loss of wet grassland, the redundancy of functional features such as ponds and the levelling of topographical variations in fields. Remaining areas have therefore become increasingly important for the rich diversity of wildlife associated with these habitats.

For birds, wet areas are a particularly good source of invertebrate food. Such areas are known to be valuable for yellow wagtails, song thrushes, starlings, reed buntings, tree sparrows and waders such as the lapwing. As wet grassland has declined, the edges of open water have become a more important nesting habitat for birds such as grasshopper warblers and reed buntings. Clean, open water habitats are also important for other vertebrates, notably otters, water voles, fish and amphibians. Many invertebrate groups are associated with water, such as dragonflies and mayflies.

A Hay (The RSPB)

Plate 3.8
Any rough vegetation next to ditches provides food and nesting habitat for a range of birds, whilst open, shallow banks provide foraging habitat for species such as lapwings.

3.4.2 Practical management

Ditches (or drainage channels)

Ditches can provide opportunities for a wide range of wildlife, while retaining their drainage function. Both seasonal and permanent wet ditches are important, each having their particular wildlife value (Plate 3.8).

Cleaning management: over time, ditches undergo natural succession. In wet ditches, tall emergent plants such as rushes and reeds shade out aquatic plants, and accelerate silting. Each phase has its characteristic wildlife. As well as impeding drainage of the surrounding land, extensive vegetation and silt reduce habitat for wildlife that require the earlier stages of succession. However, if large sections are completely cleared in one go, recolonisation will be slow and the wildlife value will be reduced. The greatest wildlife diversity is found in ditches where all the successional phases are present. This will require a little and often approach to cleaning, so that the ditch system is cleared gradually over a number of years.

In large drainage systems, a range of dredging rotations may be appropriate. The most important drainage channels could be dredged on a more regular basis than those that have less of a drainage function.

An alternative, if long stretches have to be cleared in one go, is to leave one-third of the width unexcavated in order to maintain a fringe of aquatic plants. Next time, this side can be cleared and one-third left on the other side. If this would seriously affect land drainage, consider widening small sections, perhaps at channel intersections, so that some silted and vegetated areas can remain. Weed-cutting buckets enable excessive vegetation to be removed while retaining many aquatic plants, so that the gap between full clearance operations can be extended.

Avoid cleaning during the peak nesting season between March and the end of August. Herbicides should not be used to control vegetation in ditches or on ditch banks. Ideally, a 5 m pesticide-free buffer zone should be maintained from the top of the bank.

Ditch profile: some species require particular ditch profiles. While water voles and kingfishers prefer steeper banks, the majority of wetland plants and invertebrates require very shallow water. Shallow ditch edges of less than 45° are particularly important in areas where waders breed. They provide important feeding areas for the waders and their chicks and prevent chicks being trapped in steep-sided ditches.

Re-profiling can improve ditches with poor wildlife value, but may take up slightly more land, so a compromise might be to only re-profile one side. Incorporating gently sloping stock drinking points along the channel also creates suitable habitat (Figure 3.4).

Varying the depth of the ditch prolongs the availability of open water in channels. This can be done by deepening some sections or using a bund or sluice to hold water back. Avoid straightening meanders in ditches as this will reduce the habitat available.

Ditch banks: allowing isolated trees or patches of scrub to develop can add to the habitat mosaic. Avoid planting new hedges close to the southern bank as this will shade out large sections of the ditch. Where the surrounding land holds breeding waders, avoid planting new trees or hedges and keep existing hedges short. Taller hedges and trees offer vantage points for predators and waders will be put off nesting in such areas.

Grazing of banks prevents scrub and tall vegetation building up and will slow down the rate of succession. Disturbed ditch edges provide habitat for annual plants and a variety of invertebrates. However, some sections of taller, un-grazed vegetation provide diversity, benefiting wildlife that require greater cover, such as water voles. Deciding whether to allow stock access to ditches is dependent on the likely damage to ditch banks and how efficiently the ditch needs to function as a drain or wet fence. A

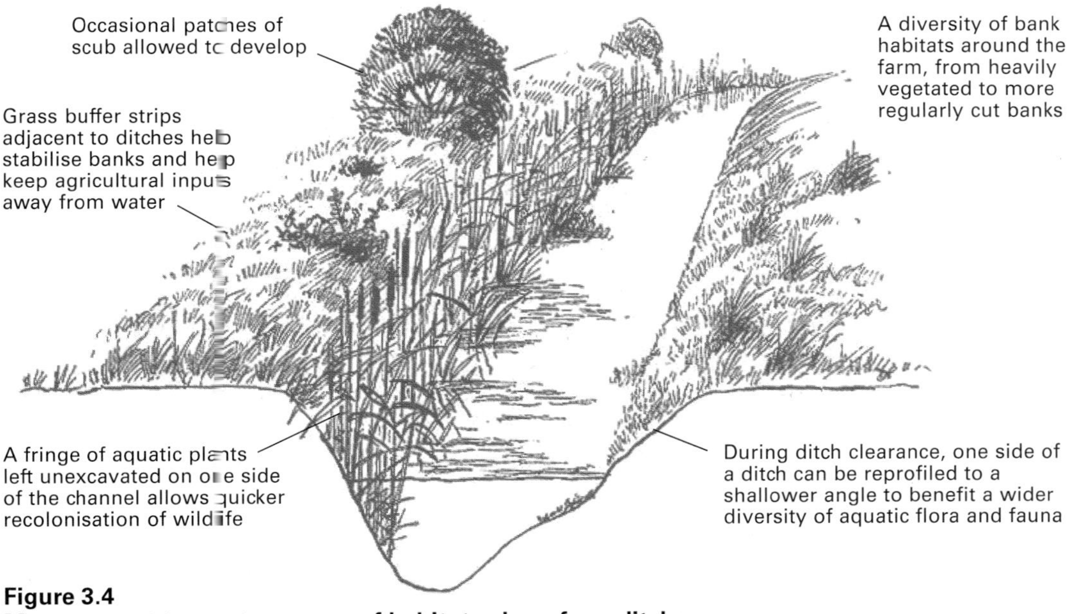

Figure 3.4
Management to create a range of habitats along farm ditches.

Ditch edges can be pulled back at intersections or along short stretches to create shallow water or exposed mud. These invertebrate-rich areas are benficial to breeding waders and other birds. Fencing off one side of a ditch can help create a variety of ditch habitats.

Figure 3.5
Methods of providing access to ditch banks and open water for birds such as waders.

compromise may be to fence off some sections of the ditch system.

In arable situations, maintain a variety of bank habitats around the farm by cutting banks on a rotation of two to five years. Regular cutting prevents a few species from becoming dominant, while tall, marginal vegetation provides nesting sites for species such as reed warblers and reed buntings. Cutting should be avoided during the breeding season between March and the end of August.

Modifications to drainage systems offer opportunities to re-create damp areas around the farm. Ditches can be widened at intersections or along short stretches to create shallow water and exposed mud (Figure 3.5). These invertebrate-rich areas are beneficial to many birds. Breeding waders need such areas to remain wet into late June, so a sluice or sandbags could be used to retain water.

Farm ponds

Although some ponds are natural features, most have been created over the centuries to provide water for livestock, to extract marl or provide fish. Over the years, many of these ponds have been colonised by a rich variety of aquatic flora and fauna (Plate 3.9). Ponds also provide feeding, drinking and bathing opportunities for a range of other species. Shallow margins warm up quickly in the spring, hosting a rich diversity of invertebrates. In turn, these are rich feeding areas for a number of birds, mammals and amphibians. Different types of ponds have value to different wildlife. Seasonally flooded pools are a particularly important habitat that supports a number of rare invertebrates.

Ponds may require occasional maintenance to ensure they do not fill up with plant debris and sediment. Only clean out around one-third of the pond in any one year to ensure that species are not lost. Livestock access to pond margins can be

Andy Hay/rspb-images.com

Plate 3.9
A combination of open water, emergent vegetation and bank-side vegetation provides a wide range of habitats for wildlife, especially aquatic invertebrates.

beneficial, keeping vegetation short and creating areas for annuals to grow. However, taller vegetation around ponds provides habitat for other wildlife. If possible, create a variety of habitats by providing stock drinking points or only fencing off part of the pond. Stocking a pond with fish reduces its wildlife value.

When creating a new pond, do not excavate sites of existing conservation value such as seasonally flooded and damp grassland. Boggy areas will often be of greater value to wildlife than a pond. Most of the pond edge should have a shallow slope and the edge can be maximised by an irregular outline. New ponds are quickly colonised by wetland plants and invertebrates. They should not need to be planted up.

Rivers and streams

Rivers and streams are extremely important wildlife habitats. The wildlife value is improved by features such as meanders, islands, deep pools and muddy margins with emergent vegetation. Predominantly, management should be about preventing pollution. Care should be taken to avoid erosion from heavy trampling and grazing along watercourses. This may be best achieved by establishing a buffer strip adjacent to the watercourse.

Water quality

The wildlife value of wet features is dependent on high water quality. This is influenced by the management of the surrounding land, including the storage and use of slurries manure, dirty water, silage effluent, fertilisers, fuel oil, pesticides and sheep dip.

Agricultural inputs need to be applied with care. A farm waste plan will help reduce the risk of water pollution. Slurry and manure should not be spread within 10 m of water bodies. Local Environment Risk Assessment For Pesticides (LERAP) and deflector plates on fertiliser spinners also help keep inputs away from water. In field, timely and appropriate application rates reduce leaching and wastage.

Buffer strips can be created to protect water bodies using agri-environment schemes or set-aside. As well as buffering ditches or watercourses from agricultural inputs, buffer strips can be managed so they are valuable wildlife habitats in their own right (see sections 3.9 and 3.24 on management of field margins in arable and grassland).

Good soil management on grass and cropped land retains valuable topsoil and minimises nutrient loss into watercourses.

Visual monitoring of water bodies can give an indication of water quality. Water heavily loaded with sediment may indicate soil erosion issues. Algal blooms, thick blanket weed and a mat of floating duckweed suggest nutrient enrichment. Water bodies containing clear water, a range of plants and abundance of insects are signs of good water quality.

Further reading

DARD (Northern Ireland) (2003) *Code of Good Agricultural Practice for the Prevention of Pollution of Water*. DARD, Belfast.

Environment Agency (2001) *Best Farming Practices: profiting from a good environment.* R & D Publication 23.

MAFF/WOAD (2000) *Code of Good Agricultural Practice for the Protection of water* PB0587. Ministry of Agriculture, Fisheries and Food.

Scottish Executive (2002) *The Prevention of Environmental Pollution From Agricultural Activity (PEPFAA) Code Dos and Don'ts Guide*. Scottish Executive, Edinburgh.

RSPB, NRA and RSNC (1994) *The New Rivers and Wildlife Handbook*. The RSPB, Sandy.

Williams P, Biggs J, Whitfield M, Thorne A, Bryant S, Fox G and Nicolet P (1999) *The Pond Book: a guide to the management and creation of ponds*. Ponds Conservation Trust, Oxford.

3.5 Farm buildings

- Though traditional buildings tend to have the greatest wildlife value, a lot can be done to improve the wildlife value of modern agricultural buildings.
- Retain and incorporate features that attract wildlife wherever possible.
- Avoid disturbing breeding or roosting birds and bats during renovation work.
- Ensure good feeding habitat is provided around the farmyard and in the surrounding farmland.

3.5.1 Wildlife use

Farm buildings are important habitats for several species of birds (Plate 3.10) and bats. Old, traditional buildings may also be important habitats for plants such as ferns, mosses and lichens, and numerous invertebrates. Features such as eaves, beams, ledges, crevices in walls and access to roof spaces provide suitable nesting and roosting places. The swallow nests in few places other than barns. Roost sites and food found in farmyards aid winter survival for a number of species. Table 3.6 summarises the features required by species most likely to nest and roost in farm buildings.

3.5.2 Practical management

Older, traditional farm buildings, made from natural materials, have the greatest wildlife value. They require maintenance to prevent them from falling into disrepair and being lost. This is increasingly achieved by finding new uses for such buildings. It is vital that wildlife is taken into consideration when traditional buildings are restored or converted. It is especially important to retain features that have attracted wildlife in the past wherever possible. Some treatments to the inside or outside of buildings are toxic to wildlife. Even something as seemingly innocuous as removing mosses and lichens from roofs and walls has detrimental impacts. Besides the wildlife value they have in their own right, many species of invertebrate will be associated with these plants. These in turn potentially provide birds and bats with food.

Local authorities have a legal responsibility to establish the presence of barn owls, bats and other protected species before building commences and may require mitigation as part of the restoration work. All 16 species of bat in the UK are fully protected, as are roost sites. A habitats regulations licence is required before any work can be undertaken on such sites.

The order of priority for maintaining nesting and roosting sites in buildings is:

- leave existing nest sites alone
- create new entrance holes
- create new opportunities in existing buildings, such as internal ledges or boxes fitted under exposed overhanging roofs
- use external nestboxes as a last resort.

New buildings tend to provide fewer opportunities for wildlife. However, much can be done to improve the value of

John Markham/rspb-images.com

Plate 3.10 Old farm buildings with access to roof spaces and ledges are good places for barn owls to nest and roost.

Table 3.6 Features in farm buildings required by species of conservation concern.

Species	Typical nest sites	Alternatives	
Barn owl 	Barn owls were so valued for rodent control that many barns were built with small windows to allow them access. Barn owls will nest on internal ledges and floors above rafters	Buildings can continue to provide nesting and roosting sites for the barn owls after conversion, through an owl window and loft. Behind the window, board off an area beneath the apex of the roof to provide a floor area of not less than 90 cm x 90 cm. A half partition allows the nesting female more protection. Alternatively, a large box inside the upper roof space of buildings or high on an external wall may be used, but suitable perches will need to be constructed where absent. Boxes can also be located on poles sited within field boundaries	
Kestrel 	Uses open access through a loft gable window or similar. Nests on internal ledges and sills, usually in proximity to window	Window and loft or large box, high on external wall	
Swallow 	Relies heavily on access to farm nesting sites as it feeds on insects over the open countryside, particularly in association with grazing livestock. Nest sites are traditional and provided the food supply has not changed, the same nests or sites are used from one year to another, frequently by the same two birds. Prefers outbuildings that provide dark ledges or rafters	Maintain access to the inside of buildings wherever possible. Swallows can enter a building through a very small gap (5 cm high x 7 cm wide) and need very little light. Maintain ledges/beams where present or provide shelving, high in the building specifically for nesting platforms	
House martin 	Nest under eaves of suitable buildings between 5–10 m above ground	House martins cannot build nests on plastic or metal surfaces, so lining eaves with suitable materials such as wooden planks will provide a nest-building opportunity	
Spotted flycatcher 	Builds a small cup nest on walls protected by climbers (eg ivy) or wall shrubs	Open-fronted box placed under a climber (eg ivy) or wall shrub	
Starling 	Nests behind soffits and facias. Also, inside chimneys and between wall cavities	Nestbox, hole 45 mm. Box to be about 30% deeper than conventional box	Internal or external nestbox. Ideally, small holes on the outside of a building giving access to a larger open space provide opportunities. Internal boxes can be located behind facias and soffits. During repointing, leave gaps with access to existing nest sights unpointed
House sparrow 	Nests behind soffits, facias and holes in brick/stone work. Also nests under roof tiles and ridges	Nestbox, hole 32 mm. Nests in loose colonies – several boxes near to each other will be of greatest benefit	
Tree sparrow 	Nests behind holes in brick/stone work and under roof tiles and ridges	Nestbox, hole 28 mm. Nests in loose colonies – several boxes near to each other will be of greatest benefit	
Bats 	Houses and buildings (both old and modern). Will roost under eaves, tiles, flashings, and in lofts, cavity walls, crevices, and subterrain sites such as cellars and icehouses	Maintain access to existing sites. If bats are present, or if there is a roost, a Habitats Regulations licence is required before any work is undertaken. This licence will require a bat survey, which will determine what can be done. Any chemical treatment to a bat roost will also require permission. Do not use surface coating or timber treatments that are toxic to wildlife. Permethrin and Cypermethrin are least harmful to mammals and any permission may specify the requirement to use such a chemical. Some species will use bat boxes, but these cannot be used to replace normal roosts	

modern buildings to birds and other wildlife. Nestboxes for a range of species can be attached to the inside or outside of farm buildings. Even buildings that need to exclude wildlife, such as grain stores, can have nestboxes attached to the outside or boxes incorporated into the building with entrance holes on the outside. Entrance holes are best faced towards undisturbed cover such as trees, but avoid planting trees too close to buildings (see below). House martins and swallows find materials such as steel unsuitable for building nests on. Lining eaves with wooden planks will provide them with the opportunity to build their own nests. Alternatively, several pre-fabricated nests may be placed alongside each other to encourage house martins. A combination of both methods can be used, as house martins have often been stimulated to build their own nest alongside an artificial one.

The area around farm buildings makes an important contribution to the wildlife value of the farm. Insects attracted to livestock and manure heaps are sought by bats, swallows, pied wagtails and other birds. Planting native climbers such as ivy to grow against buildings, and landscaping the area around buildings with trees and shrubs of local provenance provides habitat and helps blend modern buildings into the landscape. Avoid planting trees within 4 m of a building or directly blocking flight paths to nest sites of species such as barn owl, kestrel, swallow or house martin.

Encouraging rough areas of grassland, brambles and other weeds provides nectar for invertebrates as well as cover and seeds for birds. Spilt grain and hayseeds provide food for sparrows, finches and yellowhammers, aiding winter survival. Water around the yard can be especially attractive during excessively cold or hot spells. Owls and other wildlife can drown in water troughs as they try to drink. Floating a wooden plank or plastic tray in the trough will help prevent this.

Providing nest sites alone is not enough to maintain populations of birds. As well as food resources around the farmyard, good foraging habitat in the surrounding farmland is also vital. The success of birds breeding in or around farm buildings is linked to the distance they need to travel in order to find food, with higher success rates when food is nearby. For example, it is estimated that a breeding pair of barn owls require around 1.5 ha of rough grassland within 2 km of nest sites.

Less welcome mammals in farm buildings are rats. New, second generation rodenticides, such as brodifacoum, bromadiolone and difenacoum, are much more toxic to domestic animals, birds of prey and other wildlife than warfarin, so these will need to be managed with extreme care. In particular:

- prevent access to bait by birds and other animals
- search for rodent bodies and dispose of safely
- remove all remains of bait and bait containers after treatment and dispose of safely.

Where rodent infestation is a serious problem, pest control specialists should be consulted.

3.6 Farmland bird feeding stations

Feeding stations can provide important seed food for birds in areas with no overwintered stubbles or wild bird cover, or can be used to complement these and provide seed food further into the spring after stubbles have been cultivated. It is best to start feeding in October and move the feeding station along a track or hedge to reduce the risks of disease and use by rodents.

3.6.1 Wildlife use

Seed food for birds through the winter can be provided by overwinter stubbles, rotational set-aside or wild bird cover. However, on some farms, the easiest option is to put out grain or grain waste along a farm track regularly through the winter (Plate 3.11).

Many farmland bird species require seed food throughout the year to survive. This food resource is most limited during the winter months and the early spring. Feeding stations are particularly important in areas where overwinter stubbles cannot be left uncultivated for birds to feed on, and where it is not possible to establish a wild bird cover crop with seed-bearing plants to provide seed food through the winter.

The quality of overwinter stubbles and wild bird cover depends on the amount of seed food within them, and a feeding station can be used to supplement these if required (ie if the stubbles are devoid of any broad-leaved weeds or spilt grain, or if the wild bird seed mixture has failed to produce seed).

There is also an important period between February and April, when many stubble fields have been cultivated and the seed content of most wild bird covers has become depleted, but before other seed sources are available. Providing seed food for birds from February until April may be critical in terms of filling a gap in food availability and ensuring that the birds are fit enough to breed successfully.

3.6.2 Practical management

Where to site a feeding station

It is best to feed the birds regularly, so the location should be one that is visited frequently during farm work. The site should be away from grain stores and animal feed stores. It should be in an area used by farmland birds such as sparrows and yellowhammers in winter, and close to a hedge or bushes so that birds can quickly

Andy Hay/rspb-images.com

Plate 3.11 Yellowhammer feeding on grain spread along a farm track.

find cover if they are disturbed. An open site away from hedges may attract skylarks, but is also more likely to attract wood pigeons and crows.

A non-cropped area, such as a farm track or area of short grassland, would be ideal as there is no risk of re-introducing weed seeds onto arable land. A firm, free-draining surface is best as there is less risk of the seed rotting or germinating. It is important that the seed is not spread in thick vegetation, as it needs to be clearly visible to attract the birds to the site. Broadcasting seed into wild bird cover to supplement the seed food that it provides may extend the period that this habitat is used by seed-eating birds.

What feed to use

Sparrows, finches and buntings feed on a variety of seeds including grain and small weed seeds. Tailings, the by-product of the grain cleaning process, can provide useful food if there is a high content of weed seeds and chipped or small grains. Pure cereal grain can also be used, but do not use tailings that are mostly husk or chaff as they are of no benefit to the birds and just create a mess. Given the choice, birds prefer wheat to barley or oats because it has no husk.

Storage

Keep the seed dry and away from pests. It will be more convenient to store the seed near to the feeding site. Galvanised rodent-proof grain bins that are well protected from the weather are ideal.

When and how to feed

It is best to start feeding in early October so that the birds become used to a reliable food source before their other food supplies run out. Carry on feeding until at least the end of March. If possible, continue for as long as the birds are feeding at the site.

It is best to spread the seed thinly but ensure that it is clearly visible to birds. The amount of food needed will depend on how many birds are attracted, how much other food is available and the weather. The best approach is to judge how much to put out each week from the amount that was used in previous weeks. Be prepared to increase the amount put out as demand increases from October to December/January. Demand will be higher in periods of cold weather. The amount is likely to vary from 10 kg to 25 kg per week. On this basis, half a tonne of seed should be sufficient for a whole winter.

Do not wait more than a week before putting fresh food down. If there is still food left, reduce the amount slightly. It is a good idea to use an adjacent stretch of track each time so that new food does not pile up on top of old food. This will reduce the risk of diseases such as salmonella spreading through the bird populations. It will also reduce rodent use of the feeding station, as farmland birds are more able to make use of a mobile feeding station that moves along a track or hedgerow through the winter.

3.7 Wild bird cover

Wild bird cover is also known as wild bird seed mixtures in English agri-environment schemes and unharvested crops in Scottish agri-environment schemes.

A very rich food source for seed-eating birds through the winter can be established using a mix of seed-bearing crops that are not harvested. These crops can also be valuable for the conservation of arable plants and invertebrates. Different species have different seed preferences, and mixtures can be tailored to the needs of target birds.

The management of the crops should reflect good agronomy for the crops concerned, although use of pesticides is very restricted due to the importance of broad-leaved weeds and insects within the crops.

3.7.1 Wildlife use

In this section, the term wild bird cover refers to mixed crops of seed-bearing plants established primarily to provide seed food for birds, and not game cover crops, where the primary function is vegetation cover to protect game birds, although wild bird cover crops can serve both functions.

Winter food for seed-eating birds

Eleven species of red-listed birds are seed-eating birds that would be supported by winter food provision, of which five are thought to have declined due to poor overwinter survival (Table 3.7). A joint study by the Game Conservancy Trust (GCT) and the British Trust for Ornithology (BTO) showed that in winter, seed-eating birds were generally found foraging in wild bird cover more than any other habitats on farmland with the exception of winter stubbles for skylarks and grassland for grey partridges. Species differed in preferences for different seed types, and often seed preferences altered through the season as sources of certain seeds became available or depleted at different times, according to how long they were held by the plants. Seed preferences are generally related to bill shape and size,

Table 3.7 Preferences for various components of wild bird cover and location, based largely on GCT/BTO study, but supplemented by other studies, especially for species not represented in this study.

Species	Seed preference	Preferred structure and location
Grey partridge	Kale, rape, weeds	Thick canopy such as kale in open locations
Skylark	Kale, linseed, cereals, quinoa	Open canopy such as sparse cereal away from boundaries features
House sparrow	Wide range, especially cereals	Close to farm buildings
Tree sparrow	Wide range, especially kale, rape, quinoa, millet and weeds	Next to thick hedges, small woodlands or scattered trees
Bullfinch	Wide range including kale, quinoa and millet	Next to tall hedges or woodland
Linnet	Mustard, kale, linseed and rape	Next to hedges or areas of scrub
Twite	Kale, root crops and weeds	Open farmland
Cirl bunting	Cereals	Next to hedges
Yellowhammer	Cereals	Next to hedges
Reed bunting	Very wide range of small seeds eg millet, rape and weeds	Next to hedges, close to ditches
Corn bunting	Cereals	Away from boundary features other than grass strips

R Winspear (The RSPB)

Plate 3.12 For most species, the ideal location for wild bird cover is next to a thick hedge.

in relation to the size and hardiness of the seed, as well as nutrient content. The nutritive value of cereals and other grasses is lower than that of other components of wild bird covers but cereals are the staple diet of yellowhammers and corn buntings, and are taken by all the seed eaters except the small billed finches, such as linnets and twites. Weedy mixtures can be beneficial in terms of the seeds they produce and the insect life that they support. In particular, grey partridges, tree sparrows, twites and reed buntings all preferentially forage in weedier mixtures in the winter.

Crop structure and location are important considerations for many species. Skylarks' use of wild bird cover in late winter may reflect their preference for the more open crop structure at this time, whereas cover may be the most important attribute for game birds. Plots abutting the field edge are preferred by more species, and are strongly preferred by the chaffinch, greenfinch, linnet and yellowhammer. Tree sparrows, bullfinches, greenfinches and linnets prefer plots adjacent to tall hedges, whereas skylarks and corn buntings may benefit from plots established away from cover, especially trees and tall hedges.

Summer use

Wild bird cover may be used by nesting and foraging skylarks and corn buntings where plots are sited in open locations. Yellowhammers use cereal crops and cereal-based seed mixtures for summer foraging, but avoid kale-based mixtures at this time. Ripening rape in mixtures may be an important food source for linnets feeding chicks – spring and winter rape varieties provide this food source at different times through the breeding season, and therefore a mix around the farm would spread the benefits through the season. An annual mix of cereal and rape drilled at a lower seed rate in the autumn or spring can provide an ideal chick-rearing habitat for the grey partridge.

Wild bird covers generally do not provide much seed food in the spring, but this is thought to be a limiting factor for turtle doves after they have arrived in May, so any capacity to provide a seed-rich habitat for them would be very valuable. It may be that open, weedy, second-year mixtures offer the best habitat for them.

A study in Scotland showed much higher densities of birds foraging in wild bird cover crops through the summer than nearby arable crops. The abundance of weeds, butterflies and bumblebees were also much greater.

Other wildlife benefits

Song thrushes also use wild bird cover for foraging, especially brassica-based mixes where the damp understorey provides ideal conditions for feeding on molluscs and earthworms. The whole range of herbivorous mammals, some of which are regarded as pests, also use the crops. If a rodent infestation is likely to cause a problem then use of a more open crop such as an annual cereal-based mixture should reduce numbers.

The insect fauna of wild bird cover may be improved by including crops that attract pollen and nectar feeding insects such as rape and linseed, or phacelia, for which the primary function would be summer insect food rather than a seed source for birds. Weedier crops also generally support more insect life as the weeds act as food plants to a wide range of insects. Wild bird cover yielded more observations of butterflies on the Loddington Estate in Leicestershire than any other habitat. Biennial crops give a two-year break from cultivations, which can enable some weeds to flourish and insect populations to build up, such as sawflies, which require undisturbed conditions in the autumn and winter when pupae are in the soil.

3.7.2 Practical management

Wild bird covers are generally established in the spring to provide seed food for one or two winters before being renewed. In situations where spring sowing is not possible, then autumn sowing (eg a winter cereal/linseed mix) on a two-year rotation provides seed in the second winter only. Adjacent strips created in alternate years will be required to provide seed every winter. Rotation of wild bird cover prevents the build up of diseases such as brassica clubroot. In livestock areas, part of a cereal crop could be sown with linseed and left at harvest before reseeding the field in the spring. This enables livestock farmers to create a winter seed source that does not need to be fenced off from stock.

The GCT/BTO study identified kale as being the most preferred crop for the widest range of species, followed by turnips, quinoa, cereals, linseed and rape. Kale is a biennial crop (seeding in the second year) and can form the basis of a two-year mix. A kale and quinoa mix is ideal for mixtures sown in late April/May to avoid the risk of losing the quinoa in a late frost, or a kale and cereal mix can be established earlier (March/April) to allow the cereal to yield better. With the exception of turnips, the other preferred crops are annuals that may not regenerate in the second year, so they can accompany kale in a two-year mix to provide seed in the first winter, or they can be used in an annual mix. Turnips do not act as a good seed source themselves, but often act as a good nursery crop for broad-leaved weeds. A mix of two-thirds cereal and one-third linseed drilled in the autumn or early spring is ideal for yellowhammers and corn buntings as the linseed will not compete strongly with the important cereal component, but will provide an element that attracts a larger range of species including the linnet and the reed bunting. Maize, buckwheat, phacelia, sunflowers and canary grass are poor components because they are either little used by birds as a food source, or they only attract one or two species.

Crops vary in terms of when their seed becomes available and when it is depleted within the mixture. Cereal grains often rot or germinate soon after dropping to the ground, whereas the oily seeds of other species last longer on the soil surface. Species such as corn buntings feed at ground level, so sections of a standing cereal crop could be sequentially rolled or cut through the winter to give birds access to new grain. Quinoa, linseed and millet generally retain seed into the new year, whereas spring barley often lodges early in the autumn and kale seed is often depleted by late December.

Larger plots often have lower bird densities (birds/ha), but retain them for longer as the total seed supply takes longer to be depleted. The GCT/BTO study showed that plots of at least 1 ha generally retained a seed supply throughout the winter.

Weeds can be an important component of the mix depending on the spectrum. If they consist of perennial weeds, grass weeds or cleavers, then they are more harmful to the crop and less beneficial to the wildlife than if they consist of annual broad-leaved weeds. Problem weeds can be managed using a stale seed bed cultivated in February and sprayed off prior to drilling in April. Use of pesticides in the growing crop and spray drift from adjacent crops should be avoided if possible due to the importance of broad-leaved weeds and insects in the crop for birds.

The old crop can be destroyed using a coarse topper in February followed by ploughing or discing to allow breakdown and incorporation by the time that the crop needs to be re-established. Rotational management of this type would be important to ensure that some plots were available for foraging throughout the late winter.

Fertiliser needs are best determined by soil testing every four to five years. Best sowing dates will depend on the crop composition and the soil temperature and moisture. A fine, firm seedbed is required for small-seeded crops such as kale.

The most useful components of a mix are described below. Suggested seed rates per hectare of the different components are given in Table 3.8.

Kale: biennial. There is a risk of clubroot disease build-up seriously limiting yield if grown continuously. This can be overcome or reduced by rotating plots around the farm, using a non-brassica break crop, using a resistant variety (eg Caledonian kale) or ensuring an optimum pH of 6.5–8.0. Kale is a nitrogen-demanding crop, ideally receiving 70–100 kg N per ha. Flea beetle can be a significant pest. Kale can be purchased with a seed dressing to provide early protection, but this must be drilled and must not be broadcast and incorporated. Alternatively, use of a range of varieties that go through different growth stages at different times is thought to reduce the scale of attacks, and inclusion of fodder rape in the mix provides a more attractive food plant for the flea beetle, thus protecting the kale.

Cereals: annuals. Wheat, barley, oats or triticale can be used. Oats and triticale are less prone to lodging and diseases, and require less nitrogen. Oats also have smaller, softer grains that are more suitable for species such as the tree sparrow and reed bunting. Triticale is a taller, denser crop, which may make it less suitable where creating nesting habitat for buntings or skylarks is an objective. It is more resistant to rabbit grazing and holds the grain on the ear for longer, so providing food later in the winter. Establish early: autumn varieties in the autumn or spring varieties in March or early April. Where the objective is to attract nesting corn buntings, establish a spring cereal mix in areas used by nesting corn buntings in the past.

Quinoa: annual. A prolific producer of protein-rich seed attractive to a wide range of species. Best drilled late April or May to avoid late frosts.

Linseed: annual. As well as being a good seed source in itself, it is a good companion to a cereal in an annual mix as it maintains an open structure that is preferred by some species. Autumn or spring varieties are available.

Table 3.8 Suggested seed rates for crops – if part of a mixture then divide figure by the number of crops in the mix.

Crop	Seed rate kg/ha
Kale	5
Cereal	125
Quinoa	10
Linseed	60
Rape	7.5

Rape: annual. May be difficult to establish in small plots where there is a risk of grazing by wood pigeons. Prone to a number of insect pests including flea beetle during establishment, pollen beetle at bud stage (especially spring-sown varieties), and seed weevil and pod midge at pod set. Insecticides are generally not permitted on set-aside or agri-environment scheme options.

Further reading

Boatman N D, Stoate C (2002) Growing crops to provide food for seed-eating farmland birds in winter. *Aspects of Applied Biology* 67: 229–236.

Boatman N D, Stoate C, Vickery J A and Henderson I G (2001) *Designing crop/plant mixtures to provide food for seed-eating farmland birds in winter*. Report BD1606, to Department of Environment Food and Rural Affairs.

Henderson I G, Vickery J A, Carter N (in press) The use of winter bird crops by farmland birds in lowland England. *Biological Conservation*.

Parish D M B and Sotherton N W (2004) Game crops as summer habitat for farmland songbirds in Scotland. *Agriculture, Ecosystems and Environment* 104: 429–438.

Penford N and Ferguson S (2001) *Wild bird cover: a review of mixes, management and application in Scotland*. A report to the RSPB prepared by Grampian FWAG.

3.8 Pollen and nectar mixtures

Mixtures of pollen and nectar-rich flowering plants are established and managed to benefit insects by creating a continuous supply of flowering plants through the summer.

In addition, other sources of pollen and nectar produced by plants in hedgerows and field margins should be encouraged by appropriate management.

3.8.1 Wildlife use

Pollen and nectar mixtures are sown mixes of flowering plants rich in pollen and nectar, such as clovers and vetches (Plate 3.13). The objective is to provide a continuous supply of pollen and nectar for pollinating insects from March until September. Bumblebees are a particular group for which these mixes are targeted. Different species have different lengths of proboscis and therefore feed from different flowering plants, so a variety of plants is desirable.

3.8.2 Practical management

Pollen and nectar mixtures are best sited in locations that receive a lot of sunlight such as the south sides of hedgerows or along open field boundaries. The patches do not need to be large, and many small patches scattered around the farm are better than one large patch as they will ensure that the abundance of insects is increased over a larger area.

The mix should consist of a range of legumes, such as red and alsike clovers, birdsfoot trefoil, sainfoin, common vetch and black medick. Fine grasses, such as fescues, bents and crested dogstail can make up 80% of the mix to reduce the impact of weeds. Some other plant groups, particularly those capable of surviving within a grass mixture, such as black knapweed, can be added. A wider range of flowering plants will increase the flowering period of the mix as a whole. A mix of 80% fine grasses and 20% legumes should be sown at a seed rate of 15–20 kg/ha.

Glyphosate may be required to create a stale seedbed before sowing. Weed control, especially of thistles and docks, may be necessary using a weed wiper or spot treatment. The mix will not require any fertiliser, and steps should be taken to prevent the drift of any applications of pesticides, fertilisers and lime from the adjacent crop. The flower density of the

R Winspear (The RSPB)

Plate 3.13 A pollen and nectar mixture established as a 6 m margin strip.

Table 3.9 Plants with flowers attractive to bumblebees.

Species	Flowering period		
	Spring	Early summer	Late summer
Sallow	✔		
Wild cherry	✔		
Dandelion	✔		
White deadnettle*	✔	✔	✔
Common vetch*		✔	
Bush vetch*		✔	
Tufted vetch*		✔	
Hogweed		✔	
Wild roses		✔	
Hedge woundwort*		✔	
Honeysuckle*		✔	
Red campion*		✔	✔
Foxglove*		✔	✔
Bramble		✔	✔
Common figwort*			✔
Common mallow			✔
Great willowherb			✔
Greater knapweed			✔
Lesser knapweed			✔
Black horehound*			✔
Spear thistle*			✔
Teasel*			✔
Lesser burdock			✔

*Flowers attractive to bumblebees with a long proboscis

crop will decline over time and it is likely that the mixture will need to be re-established about every five years. Annual cutting and removing of the vegetation and prevention of herbicide or fertiliser drift can boost the lifespan of the nectar-bearing plants.

To boost late flowering, half of the area (the same half every year) should be cut before the end of June. It may be wise to walk through the plot prior to cutting to see if any ground-nesting birds or hares are flushed. Avoid cutting at this time of year if there is any evidence of nesting birds. If this has been an issue in previous years, then it would be advisable to mow half of the plot regularly up until late June to keep it too short to attract nesting birds. The whole area should be mown after 15 September. Ideally, the cuttings should be removed to protect the sward underneath. In grassland situations, winter grazing (September to March) is acceptable provided the ground is not poached.

Whilst pollen and nectar mixtures provide a rich food source for invertebrates, it is also important to protect other pollen and nectar sources on the farm, as the range of insects that can be supported will be related to the diversity of flowering plants. Table 3.9 shows some of the plants likely to support pollinating insects at different times of the year.

3.9 Arable field margins (buffer strips) and field corners

Tussocky grass margins can be created around arable fields to boost overwintering insect populations and provide nesting habitat for birds and mammals, or wild flower margins can be created to boost pollinating insect populations. Similar management of field corners benefits similar species, but does not provide such an effective network of habitat allowing movement of less mobile species. Natural regeneration will have the benefit of creating a sward of local provenance plants, but there is a risk of weed control problems. If a mix is sown then native plants should be used to boost insect benefits. Control of competitive weeds is the priority governing management in the first year.

P Thompson (The Game Conservancy)

Plate 3.14
A tussocky grass margin provides good habitat for a range of beneficial invertebrates through the winter.

R Winspear (The RSPB)

Plate 3.15
A flower-rich margin provides a pollen and nectar source in the spring.

3.9.1 Wildlife use

There are generally two types of grass margins or buffer strips adopted on arable farmland – tussocky grass margins composed of a mix of grasses including at least one tussock-forming species (Plate 3.14), and wild flower margins composed of a mix of fine grasses and wild flowers (Plate 3.15).

The main wildlife benefit of tussocky grass margins around arable fields is the large invertebrate populations that they support. Tussocky grassland harbours very large populations of spiders and insects as it provides protection through the winter and relatively undisturbed habitat that allows populations to build up through the summer. The invertebrates support populations of insectivorous birds and mammals and provide a source of insects for the chicks of many seed-eating farmland birds.

Tussocky grass margins around arable fields also provide nesting habitat for game birds and small mammals such as mice, voles and shrews. These in turn attracts hunting birds of prey and owls.

Wide grass margins adjacent to a thick hedge provides ideal nesting habitat for yellowhammers and whitethroats. Yellowhammers may nest in the grass at the base of the hedge early in the season before the hedge is in full leaf. However, the link with wide grass margins is more likely to be due to the need for a rich source of insect food for the chicks close to the nest site. Corn buntings also tend to nest in arable areas with wide grass margins, probably for the same reason.

Wild flower margins provide less cover for nesting birds, mammals and overwintering insects, but provide pollen and nectar sources for pollinating insects. The pollen and nectar mixtures (see section 3.8) are managed specifically for this purpose, but incorporation of some flowering plants within a grass margin is covered in this section.

To a lesser degree, grass margins also provide seed food for birds and mammals. However, access to the food resources of the margins can be limited for some bird species that cannot easily forage in the thick vegetation. This issue is discussed in the practical management section below.

Permanent grass margins are not appropriate where remnant populations of rare arable plants occur in the crop edge. These depend on regular cultivation and will be lost by establishing a permanent grass sward. Uncropped cultivated margins are designed to conserve these species (see section 3.11).

Similar management of field corners will benefit similar species, but does not provide such an effective network of habitat allowing movement of less mobile species, such as invertebrates and small mammals, around the farm.

There are many other benefits of grass margins.

- They act as buffer strips preventing spread of pesticides and fertilisers onto sensitive habitats, notably watercourses – 5 m buffer strips fully meet the requirements of the Local Environmental Risk Assessment for Pesticides (LERAPs).
- They can be used to protect the roots of hedges and hedgerow trees.
- They can reduce levels of soil erosion and runoff.
- They form a barrier to weeds spreading from the hedge-base to the crop edge (barren brome can shed seed up to 1 m from the parent plant, therefore a 1 m strip is not an effective weed barrier).
- They enhance populations of beneficial insects that overwinter in the margins (especially in tussocky grass swards) and act as crop pollinators or predators of crop pests in the spring.
- Grass margins are funded under agri-environment schemes in all countries of the UK.
- Wide strips provide easy access for hedge-trimming and allow hedge management in late winter when the impact on wildlife is minimised, although the margins should not generally be used as access tracks if wildlife conservation is the objective.
- They present a good public image of farming.

In addition, grass margins occupy the least productive area of the arable field where factors such as shade, less efficient input coverage and rabbit grazing generally reduce yield by about 20% anyway. The benefits that accrue usually more than compensate for the loss of cropped area.

3.9.2 Practical management

Existing margins

Maintaining existing margins and field corners is at least as important as establishing new ones. Special attention should be given to margins connecting semi-natural habitats on the farm or containing a diverse range of flowering plants as these are likely to be more important than any recently established margins. The best management will normally be to continue the management already undertaken, but with special effort to avoid encroachment of the hedge onto a floristically-rich margin. These margins are best managed by cutting at least once per year to prevent the grasses from out-competing the broad-leaved plants.

Establishment

Grass margins can be established in the spring or autumn, although September is ideal as the soils are warm and moist. If it is necessary to establish them in the spring then a higher seed rate is recommended. If the margin is heavily shaded (eg on the north side of a hedge) then establishment of a perennial strip may be difficult. Consider doing this following hedgerow management such as trimming, coppicing or laying.

From a wildlife perspective, natural regeneration is better than sowing a mix in terms of developing native vegetation of local provenance, but poses a potential risk in terms of weed control problems. Natural regeneration may be an option on light soils where the adjacent flora is diverse and no highly competitive weeds occur. If possible, try natural regeneration in the first instance, but if problem weeds arise then spray them off in the following spring and establish a sown mixture in the following autumn (this approach would need approval if done under an agri-environment scheme). The mix should consist of perennial, non-invasive grasses including tussock-forming species such as cocksfoot, Yorkshire fog or timothy. A good sward for predatory insects can be created using 10–20% of cocksfoot or timothy mixed with finer grasses such as fescues, bents and meadow grasses (Table 3.10). If the proportion of tussocky grasses in the mix is too high then they become increasingly dominant in the sward, and the overall diversity of the sward is lost. To some degree, species can be selected that are likely to do better in the conditions. For example, wavy hair-grass (*Deschampsia flexuosa*) is better suited to acidic soils and meadow barley (*Hordeum secalinum*) is a species of moist soils. All species in the mix should be native grasses and ideally of local provenance, especially if wild flowers are included in the mix. Agriculturally improved grasses such as perennial ryegrass (*Lolium perenne*) has very low wildlife value compared with native grasses and should be avoided.

Wild flowers can be included in the mix to benefit nectar-feeding insects, although these increase the cost and tend to do better on lighter soils. If wild flowers are to be included then they should form between 5% and 20% of the mix by weight, and the remainder should be made up of finer grasses, as tussock-forming grasses will out-compete the flower component. Species such as scabious, knapweed, yarrow, sorrel, self-heal and birdsfoot trefoil can survive within a grass margin. They are best sited along margins that receive a lot of sunlight to attract plenty of insect activity. Annual mowing in the autumn should maintain the flower component although another cut in March may be necessary if grass growth threatens to out-compete the flowering plants. It will help the wild flower component if the cuttings can be removed.

Table 3.10 Some grass species suitable for sowing in field margins.

Common bent	Meadow foxtail
Sweet vernal-grass	Crested dogstail
Sheep's fescue	Meadow fescue
Meadow brome	Meadow barley
Tall fescue	Red fescue
Smooth meadow-grass	Yellow oat-grass
Smaller catstail	Cocksfoot*
Wavy hair-grass	Rough meadow-grass
Timothy*	Quaking grass

* Tussock-forming grasses

If high populations of weeds are present then spray off with a total herbicide before cultivation. The seedbed can then be prepared with the rest of the field. Ploughing is not essential, but a fine tilth is needed. Seed can be broadcast or drilled but should not be more than 2 cm deep, and the ground rolled to get good seed–soil contact. If annual grass weeds are likely to be a problem, then a mix dominated by fescues should be used and a low rate of fluazifop-p-butyl can be applied when the sown grasses have reached three tillers. This will hit the annual grasses, but the fescues are relatively tolerant and the broad-leaved plants will be unaffected. The degree to which other perennial grasses will be affected will depend on the species, application rate and timing.

A sterile strip between the field margin and the crop is preferred by some as a means of controlling weeds at the crop edge although, if the grass margin is established successfully, then there should be no weeds available to move into the crop. If deemed necessary, then a 1 m sterile strip can be maintained by spraying or rotovating. If spraying, then a skirt on the end of the sprayer down to ground level should be used to prevent spray drifting onto the grass margin.

Three cuts may be necessary in the first summer (when the sward reaches 10 cm in height) to encourage the grasses to tiller and help control invasive annual weeds.

Cutting when problem weeds are coming into flower is necessary to achieve a perennial sward if no other cuts are undertaken. Cuts between April and August may destroy birds' nests – this can be avoided either by not cutting during this period or by cutting monthly to discourage birds from nesting (in the first year only).

Management

After establishment, the tussocky grass margins should be cut no more than once every two to three years to allow mature tussocks to form, and then only if needed to prevent scrub invasion from the hedgerow. Avoid cutting all the margins in the same year as this will have a greater impact on overwintering insect numbers and leave no tussocky swards for nesting game birds in the following spring. The margins are generally managed by cutting to a height of no less than 5 cm. If cuttings can be removed using a forage harvester then this will reduce the fertility of the strip and is likely to enhance the plant diversity in the longer term. If the cuttings cannot be removed, then a flail mower is better than a reciprocating blade or circular blade cutter as the latter leave swaths of cuttings that will kill the vegetation underneath.

One or two metre margins are sufficient to provide nesting cover for game birds. Wider margins will provide more invertebrate food and habitat for small mammals. Where 6 m margins are created, cutting the 3 m abutting the crop edge annually in the autumn but only cutting the hedge-side margin every three years will create a useful diversity of grassland structure. The tussocky margin next to the hedge may contain greater numbers of insects, but the more open sward next to the crop will be easier foraging habitat for insectivorous birds.

Wild flower margins should be cut annually, either in the autumn (may be sufficient on lighter soils) or twice – once before 15 March and again after harvest to maintain a more open grass sward that does not compete heavily with the wild flower component. If the grass sward is still too dominant then regular cutting through the summer should encourage the wild flower component and should also prevent a thick cover forming that might attract ground-nesting birds, which would be vulnerable to cuts in the spring and summer. This management may be more appropriate along regularly disturbed margins where ground-nesting birds are less likely to be attracted. Some birds may prefer to forage in the shorter grass sward as it gives them more access to the ground. It is still good practice to leave at least 1 m uncut against the hedge to add diversity and a refuge for species of undisturbed grassland.

It is important to prevent drift of non-selective herbicides into the grass margin as this destroys the perennial cover and creates ideal conditions for the establishment of annual weeds. It is equally important to prevent any fertiliser input as this will benefit the more competitive arable weeds such as cleavers at the expense of the grass seed mix and will reduce the diversity of the final sward.

Spot treatment or topping may be required to control some weeds such as creeping thistle. Thistle control should be done whenever thistles occur, before numbers are allowed to build up.

Ploughing of headlands may be important to control invasive grasses. A sterile strip is likely to do more harm than good, as it will increase the risk of annual weeds invading the crop by providing cultivated bare ground for germination immediately adjacent to the crop edge. If a sterile strip is used to control weeds then it should be positioned between the grass margin and the crop.

If the rotation includes grass leys then the margins should remain uncultivated, uncut and receive no fertiliser. If it is practical, then excluding grazing from some of the margins will maintain more of the insect populations that have built up.

Further reading

Marshall E J P (1998) *Guidelines for the siting, establishment and management of arable field margins, beetle banks, cereal conservation headlands and wildlife seed mixtures*. IACR-Long Ashton, Bristol.

3.10 Beetle banks

A beetle bank is a grass strip, normally created on a raised bank, through the middle of a large arable field to create an overwintering habitat for invertebrates, particularly predators of crop pests. Beetle banks can lie inside the cropped headland, so that the field is still managed as one unit, or connect to the field margin at each end, effectively creating a new field boundary. Grass margins on each side of the bank can be used to create wider features, boosting the wildlife benefits.

3.10.1 Wildlife use

As well as providing habitat for large numbers of overwintering insects and spiders, beetle banks can also provide good nesting habitat for grey partridges and potentially, depending on the sward structure, for skylarks, reed buntings and corn buntings. They also provide insect-rich foraging habitat for birds, particularly corn buntings and skylarks that are less likely to forage close to hedgerows. They may be used as a feeding habitat and cover for lapwing chicks

The tussocky grass sward of the beetle bank can also provide nesting habitat for small mammals including harvest mice, and cover for brown hares.

In terms of insect and spider abundance, populations may take several years to build up. The greatest abundance of wintering insects is found in the tussocks, which take a few years to develop.

3.10.2 Practical management

Establishment

Beetle banks are ideal features to establish within arable fields larger than 16 ha and wider than 400 m. Spiders and insects such as ground beetles overwinter in field margins and move into the crop in the spring. They are generally able to cover areas up to 200–250 m from the crop edge. Therefore, fields wider than 400 m tend to have low densities of these invertebrates in their centre, and beetle banks provide an overwintering habitat through the middle of such fields to fill this gap. A good principle is to divide the field into 10 ha blocks using beetle banks, so that fields of 20 ha require one beetle bank, fields of 30 ha require two banks, and so on.

The beetle bank can be established within the crop headland, so that the field is still cropped and managed as one unit with a slightly larger turning circle in the middle (Figure 3.6). It is thought that predation rate of nests along beetle banks is lower if the bank is not connected to the field margin at each end because mammalian predators are less likely to travel along them. However, beetle banks may also be used to create new field boundaries (Plate 3.16). For example, boundaries within large fields may be desirable to split areas of different soil types so that they can be managed

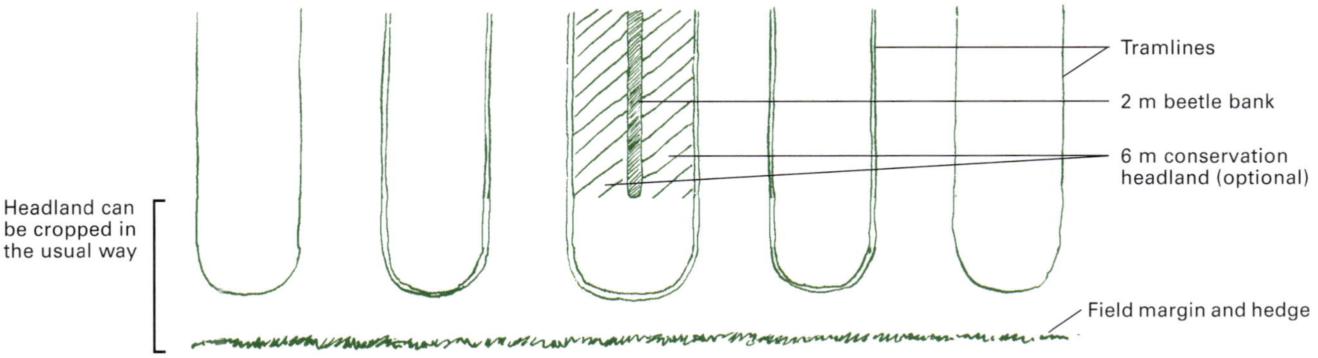

Figure 3.6
Plan view of a 2 m in-field beetle bank.

R Winspear (The RSPB)

Plate 3.16
In this case, the beetle bank has been used to split the field, with different crops grown on either side.

separately to remove the problem of uneven crop development. These banks may be used by less mobile species of invertebrates that benefit from the connection of the bank to the existing field margins.

Beetle banks are best created in the autumn when the warmth and moisture allow good grass establishment, or can be established in the spring if necessary. The bank is formed by ploughing inwards from either side during post-harvest cultivations to create a ridge 2 m wide and roughly 0.4 m high (Figure 3.7).

The important component of the beetle bank is the tussocky grasses, such as cocksfoot and timothy, as these accommodate the overwintering insects. The very low-key management after establishment also means that these are likely to dominate over finer grasses in time anyway. A high percentage of cocksfoot in the initial mix will lower the cost of the mix. However, if establishing field margins as well as beetle banks, a lower percentage of tussocky grasses should be used, particularly for field margins of 4 m or more in order to maintain a more varied sward (see section 3.9). Use of native grasses will boost the diversity and abundance of the insect life in the bank. The seed should be broadcast at a rate of roughly 70 kg/ha immediately after cultivation in order to get good establishment.

Within the first year, the priority is good establishment and weed control. Management is largely undertaken by regular cutting (in the first year only) to top the weeds and encourage tillering of the perennial grasses. Otherwise, selective herbicides or spot spraying can be used to remove unwanted weeds. Provided that a good perennial sward is established in the first year, then subsequent cutting is required no more than once every three to five years, and then only if necessary to prevent encroachment of woody species. One metre sterile strips each side of the bank may be used to prevent weed invasion of the crop from the bank, although this is only likely to be beneficial if it was not possible to establish a good, weed-free, perennial sward.

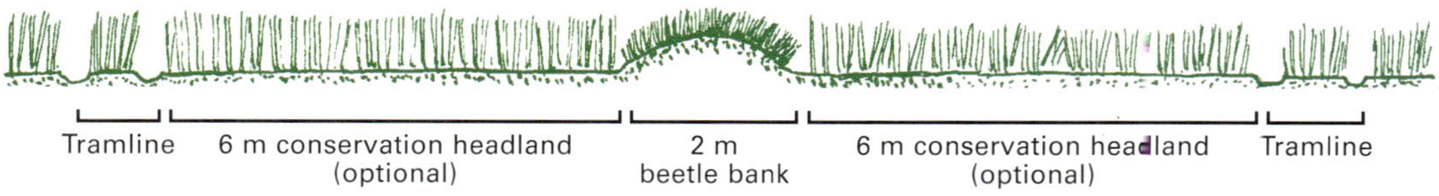

Figure 3.7
Profile diagram of a 2 m beetle bank.

Two metre beetle banks are very vulnerable to pesticide drift, although the tussocky nature of the sward will normally protect most of its inhabitants. Wider banks created by having adjacent grass margins (or buffer strips) will provide a larger invertebrate resource and be better protected from spray drift. For example, the 2 m bank can have 6 m grass margins each side to create a 14 m wide strip.

Management

Management of the beetle bank after the first year should be limited to as little cutting as possible. Cutting may be desirable if there is an ongoing weed problem, if the dead material of the grass tussocks is becoming too dominant (although some dead material is beneficial for nesting grey partridges) or if shrubs are starting to establish themselves along the bank.

Colonisation of the bank by beneficial insects and spiders is largely dependent on the pesticide regime of the field. Avoidance of broad-spectrum insecticides in the summer and use of spray buffer zones (at least 6 m) each side of the beetle bank should ensure good insect populations. On lighter soils with a low burden of competitive grass weeds and cleavers, it may be possible to establish conservation headlands each side of the beetle bank when the field is growing cereal crops (see section 3.12). This combination will provide ideal nesting and feeding habitat for grey partridges, and may boost productivity of other farmland birds. One might consider planting wild bird cover in a 6 m strip alongside the bank if a conservation headland is not used, or when the adjacent crop is not a cereal. This will greatly benefit species such as the grey partridge and probably the corn bunting and skylark, which avoid tall field boundaries.

3.11 Cultivated margins

- Arable field margins that are cultivated but not cropped provide opportunities for less competitive arable plants to flourish, and these support important insect populations and provide food for birds.
- Suitable sites are often on light or shallow soils that have a long history of arable cropping but have not received intensive inputs of herbicides or fertilisers.

3.11.1 Wildlife use

Cultivated uncropped margins are designed to enable the development of a species-rich arable flora (Plate 3.17). This measure is usually targeted at key sites for the conservation of rare arable plants.

The plants of arable land include many such as cornflower, corn buttercup and pheasant's eye that are now among the most endangered in Britain. These have declined rapidly since the mid-20th century in response to the revolution in agricultural practice that has included the adoption of herbicides for weed-control, large increases in the levels of nitrogen applied, the development of highly competitive crop varieties, mechanisation and the enlargement of fields. This former rich diversity has been replaced by a few species such as black grass, cleavers and barren brome that are highly suited to modern arable systems.

Although many of these species are now very rare, small colonies still survive in field corners, and even where plants have not been seen for many years they may persist as dormant buried seed.

These species are generally found within the outer four metres of cropped fields that have a long history of arable cropping. Cultivated margins have been shown to result in significantly greater benefits to uncommon plants than conservation headlands. Permanent grass margins are a potential threat to these dwindling populations of poorly competitive annual plants that depend on regular disturbance. This is also a measure that may be adopted on suitable soil types to encourage populations of less competitive arable plants, which are food plants for a range of insects and a potential seed resource for farmland birds and mammals. Cultivated margins also have the advantage over conservation headlands in that they allow easy access to the ground for species such as turtle doves to feed, although the cover provided by the cereal crop in a conservation headland is important for grey partridge broods.

P Wilson (Wessex Environmental Associates)

Plate 3.17 A six-metre cultivated margin consisting of broad-leaved annual plants.

3.11.2 Practical management

Information on the presence of rare arable plants is key to selecting the right sites. This can be obtained from your local Defra office or your county Biological Records Centre. A specialist survey in June or July is recommended to help locate suitable areas.

In general, the best sites are likely to be those that:

- Have been in arable cultivation for at least 100 years
- Have not had repeated, heavy applications of herbicides or fertilisers
- Are on shallow soils or light soil types (eg chalk or sand)

- Are in open, sunny locations, especially on south-facing slopes
- Are not infested with nitrogen-demanding weeds such as cleavers, barren brome or black grass.

Where there are no records for rare plant species, a combination of the above factors can be used to predict likely areas. It should be noted that there are exceptions to these guidelines, in particular, heavy clay soils in areas such as mid-Somerset, Suffolk and the West Midlands can support very rich floras.

Species of rare arable plants vary in their susceptibility to herbicides, fertilisers and crop competition, and in their requirements for soil type, water level, and timing of cultivation. Specialist advice on management should be sought wherever these plants are known to occur.

The field margin refers to the outermost 6m-wide part of the field, measured from the edge of the strip of perennial vegetation in the hedge bottom or ditch bank.

The normal management is to cultivate the field margin annually to a depth of about 15 cm (ie normal ploughing depth). Cultivations can be done with the rest of the field, but the margin is then left undrilled. However, if the rare arable plant interest is specifically for spring or autumn germinating plants, then the cultivation needs to be done in the correct season, regardless of the cultivations required for the crop in the rest of the field. In some cases, especially on very light soils, cultivations do not need to be annual, and can be carried out every two years to maintain an annual flora without encountering significant problems with perennial plants. However, the best management to reduce the problem of perennial weed infestations (thistles, couch grass etc), is to maintain annual ploughing and vary the season of cultivation. In exceptional cases, problem weeds can be removed using a spot treatment or weed wiper.

No herbicide, fungicide, nitrogen or any other agrochemical or fertiliser should be applied to cultivated margins. Cultivated margins should not receive any herbicide or fertiliser drift. Where possible, the field margin should not be disturbed until the whole field is ploughed for the following crop.

If the margins do not develop any significant wildlife interest and cause agronomic problems then it may be advisable to convert them into permanent grass margins by spraying off the weeds and sowing a grass mix (see section 3.9).

Further reading

Wilson P and King M (2003) *Arable Plants: a field guide*. English Nature/Wildguides.

3.12 Conservation headlands

Conservation headlands is the name given to cereal crop headlands with restricted pesticide inputs to allow broad-leaved arable plants and their associated insects to flourish, but still allowing control of grass weeds and cleavers, and with no restrictions on the use of fungicides or plant growth regulators. An additional step of using no fertiliser in the headlands will create a more open crop structure and further benefit the less competitive weeds. This may also reduce the burden of the more competitive weeds in the conservation headlands.

3.12.1 Wildlife use

Conservation headlands are the outer margins of cereal crops (usually between 6 m and 24 m wide) that do not receive insecticides after 15 March or broad-spectrum herbicides so that broad-leaved weeds and their associated insects can flourish (Plate 3.18). In some cases, no fertiliser is applied to the headland.

The restrictions on pesticide use on conservation headlands are designed to control invasive grass weeds and cleavers, but allow most other broad-leaved weeds to survive. These weeds may provide seed food for birds, but also act as food plants for a wide range of insects that are important food for the chicks of many declining bird species. Conservation headlands are particularly well-suited to the conservation of grey partridges because the chicks need to feed themselves from the moment that they hatch. These headlands provide a rich, accessible insect food resource immediately adjacent to the nesting habitat in the field margin. A cereal crop provides an ideal structure allowing grey partridge chicks to forage within it, but with a canopy protecting them from weather and predators. One study showed that conservation headlands increased average chick survival from 4.8 to 7.3 birds per brood. Rare arable plants may benefit from conservation headland management, especially if no fertiliser is applied.

Plate 3.18 The ideal result of conservation headland management is a sprinkling of broad-leaved weeds in the crop base.

P Thompson (The Game Conservancy)

The stubble from a conservation headland will also be a valuable source of weed seeds for birds if it can be left through the winter. Conservation headlands also act as a buffer zone, protecting adjacent field margin habitats from drift of the most damaging pesticides.

3.12.2 Practical management

Conservation headlands are most suited to light soils with no heavy burden of competitive weeds such as barren brome or cleavers. Although the herbicide regime is designed to control these, it is often less effective than conventional crop management. There is an element of risk in adopting conservation headlands on sites where grass weed build-up is likely to occur. In these situations, management that is likely to exacerbate the grass weed problem, such as minimum tillage or direct drilling, should be carefully considered before combining with conservation headland management.

The ideal location of conservation headland management is in cereal fields adjacent to good field margin (Figure 3.8) or beetle bank nesting cover for grey partridges. This enables birds to move chicks quickly from the nest site to the foraging areas after hatching.

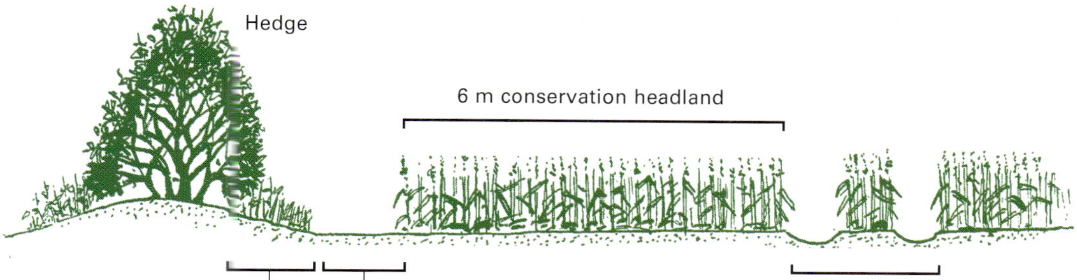

Figure 3.8
A combination of a thick hedgerow, tussocky buffer strip and adjacent conservation headland provides ideal nesting cover and insect-rich foraging habitat for a range of farmland birds.

The pesticide regime of conservation headlands is generally to avoid broad-spectrum herbicides, and only use graminicides to control grass weeds and amidosulfuron for the control of cleavers. Insecticides are not applied on conservation headlands after 15 March. Insecticides applied earlier in the crop year have less effect on insect abundance as most beneficial insects are within the field margin at this time. There are no restrictions on the use of fungicides or growth regulators. The exact list of herbicides permitted on conservation headlands has changed in recent years due to the loss of some of the key products from the approved list of pesticides. The herbicides permitted on conservation headlands that had approval in 2004 were:

- Tri-allate
- Tralkoxydim
- Diclofop-methyl
- Clodinafop-propargyl
- Flamprop-m-isopropyl.

A new range of herbicides suited to conservation headland management is currently being discussed and advice should be sought from agri-environment scheme staff on the latest information. The standard steps to prevent drift of pesticides should be taken to prevent pesticides from the main area of the crop spreading onto the conservation headlands.

Some farmers adopt conservation headlands with zero fertiliser applications. In some cases, this reduces the problematic weed burden that would otherwise develop. However, in other situations, the residual nitrogen in the soil would be enough to create a weed problem in situations where it is likely to occur. Another reason for combining conservation headlands with zero fertiliser is the conservation benefit. Many of the rare arable plants are not able to tolerate the high nutrient levels of conventional arable land, or the competition of the crop under these conditions. Conservation headlands without fertiliser applications should encourage these species. The more open crop also allows greater accessibility to the ground for birds to feed. The turtle dove in particular should benefit from this as it is a large seed-eating bird looking for seed-rich areas throughout the summer.

The crop taken from the conservation headland is likely to contain a greater level of weed-seed contamination than the crop from the rest of the field. On some estates, the headland is harvested separately and the grain used to feed game birds. In some agri-environment schemes, there are payments to leave the conservation headlands unharvested to act as a wild bird cover. The stubble of the headland should hold more broad-leaved weeds and will be a particularly useful feeding area for a wide range of seed-eating birds if it can be left through the winter to be followed by set-aside or a spring crop.

Conservation headland management is rotated around the farm with the cereal crop. If weed problems arise in particular locations then they should be recorded so that these headlands are not included again in future years.

No-spray zones for insecticides around arable field headlands should always be considered as a means of reducing spray drift into field margin habitats and leaving a refuge for in-crop insects whenever it is practical to do so.

3.13 Low-input crop management

Low-input crop management often refers to conservation headland management undertaken on the whole crop. The objective of low-input crop management is to control the most competitive weeds and damaging pests and diseases with minimal impact on non-target species of plants and invertebrates. Integrated Farm Management (IFM) and organic farming will also benefit wildlife if this objective is met.

3.13.1 Wildlife use

Some agri-environment schemes fund low-input cereals, where the pesticide restrictions are similar to those of conservation headlands, but apply to the whole crop rather than just the headland. The benefits for wildlife are similar to those of conservation headlands, but on a larger scale (see section 3.12). The crop is a particularly valuable insect food source for birds feeding chicks.

Organic crop management can also boost the abundance of broad-leaved plants and insects within the crop. Integrated farm management works on the principle of minimum use of inputs to produce optimum yields with minimum environmental impacts. If this boosts the abundance of less competitive weeds and non-pest invertebrates, then this measure will also benefit farmland wildlife.

Many of the broad-leaved weeds associated with arable land are of conservation importance in their own right. They also support a wide range of insects that boost the whole invertebrate food chain within the crop. Arable crops with high insect abundance attract a wide range of insectivorous birds and seed-eating birds that feed their chicks on insects. The broad-leaved weeds also provide seed food for birds, which is particularly important if the following stubble can be left through the winter.

3.13.2 Practical management

The idea behind low-input crop management is to control the important weeds, pests and diseases, but leave as many of the less competitive weeds and non-target insects as possible. Not using spring or summer insecticides protects the insect life that moves into the crop in the spring. Use of herbicides that selectively take out grass weeds and cleavers ensures that other broad-leaved weeds are retained to act as food plants for insects and produce seed food for birds.

Where low-input crops are funded by agri-environment schemes, refer to the prescription, as conditions vary in different countries. In terms of pesticide use on low-input cereals, adoption of the pesticide restrictions for conservation headlands (see section 3.12) will create ideal crops for a range of wildlife. Being a whole-field option, low-input cereals can be expected to contain more weeds, and are not appropriate where low weed contamination of the grain is required if there are no on-farm cleaning facilities. It may be more acceptable for cereals grown for livestock feed. In some agri-environment schemes, restrictions on seed rate and fertiliser use apply. The shorter and more open sward is more beneficial for many species foraging for insects in the crop.

Spring cereal crops offer better wildlife benefits as they can normally be grown with no insecticide sprays and many of the spring germinating weeds, such as polygonums, are very beneficial for birds and insects. They also have a crop structure that is suitable for a number of ground-nesting birds, such as skylarks, so that birds are able to nest and feed within the crop. Low seed (100 kg per hectare) and fertiliser rates (50 kg per hectare) will create a more open crop that improves the nesting and feeding opportunities. Where overwinter stubbles can follow low-input crop management, the broad-leaved weeds should provide plenty of seed food for birds through the winter.

Research has shown that organic crop management also usually results in a greater abundance of broad-leaved weeds and insects. The IFM approach to use on crops should benefit wildlife if it can result in fewer sprays, use of more narrow-spectrum chemicals and better protection of uncropped habitats. The important principles of IFM that benefit wildlife are as follows:

- Take professional advice on pesticide use from a BASIS qualified advisor.
- Use crop rotations to reduce the build-up of problematic weeds, pests and diseases in order to reduce the reliance on pesticides.
- Use crop varieties that are resistant to pests and diseases likely to be encountered.
- Use seed treatments where available if significant damage is expected, but take particular care to avoid spillages and ensure treated seed is properly drilled to prevent wildlife feeding on the toxic dressing.
- Monitor crops regularly and base control decisions on the results – treat only when economic thresholds have been exceeded.
- Where possible, use products targeted to the problem with minimum impact on non-target species; in particular, avoid broad-spectrum insecticide use in the spring and summer wherever possible.
- Apply treatments at the correct time.
- Apply in the right conditions and accurately to the area needing treatment; spray drift is related to wind speed (do not spray in wind speeds above force 3), height of boom (keep to a minimum) and spray droplet size (larger droplets are less prone to drift).
- Use buffer zones to protect sensitive wildlife habitats and watercourses.

The weeds that are important to birds as seed food and those important as food plants for insects on arable farmland tend not to be the most competitive weeds, but are the ones that are killed by herbicides intended to remove the more competitive weeds (Table 3.11). If herbicide regimes can be developed that remove the competitive weeds but leave the valuable ones, then significant wildlife benefits will accrue.

Table 3.11 The tolerance threshold of common arable weeds (seeds per square metre resulting in 5% yield loss) and their importance for birds and insects (number of associated species).

Weed species	Competitive index	Number of insect species	Importance in bird diet
Cleavers	1.7	30	
Wild oats	5	<10	
Black grass	12.5	<10	
Charlock	12.5	38	**
Mayweed	12.5	30	
Chickweed	25	70	***
Fat hen	25	30	***
Annual meadow-grass	50	52	**
Knotgrass	50	60	***
Sow thistle	50	28	
Field speedwell	62.5	<10	
Groundsel	83	47	**
Field pansy	250	<10	**

*** Important in the diet of nine or more farmland bird species
** Important in the diet of three to eight farmland bird species
Bold species indicate important species for insect diversity or key components of the diet of seed-eating birds

Further reading

HGCA (2002) *Arable Cropping and the Environment – a Guide*. Home-grown Cereals Authority, London.

LEAF (2000) *The LEAF Handbook for Integrated Farm Management: a practical guide for the adoption of integrated farming*. LEAF, Warwickshire.

Marshall E J P, Brown V K, Boatman N D, Lutman P J W, Squires G R and Ward L K (2003) The role of weeds in supporting biological diversity within cornfields. *Weed Research* 43(2): 77–89.

Soil Association (2002) *Improving biodiversity on organic farms*. Soil Association Technical Guides series.

3.14 Overwinter stubbles

Management of overwinter stubbles can be important for the conservation of seed-eating birds and scarce arable plants, and can support high numbers of some insects. The value largely depends on the intensity of weed control in the preceding crop, and can be boosted by avoiding pre-harvest herbicides and post-harvest management. The value of stubbles may be boosted by light scarifiction.

3.14.1 Wildlife use

Stubbles that are left right through the winter provide habitat for arable plants, seed food for birds and protection from cultivation for some insects.

Some weeds that germinated in the preceding crop can complete their life cycle and seed after harvest. The seeds of these weeds and any spilt grain from the harvest provide seed food for birds through the winter. The benefits for arable plants and birds will depend on the intensity of weed control in the preceding crop (Plate 3.19). Barley stubbles generally attract more seed-eating birds through the winter than wheat stubbles. This may be the result of fewer inputs resulting in more weed seeds in the stubble, and a slightly sparser sward allowing easier access to the ground. Spring cereal stubbles are particularly beneficial because of the value of some spring-germinating weeds for birds, particularly polygonums.

One of the key insects in the diet of farmland birds is the sawfly, which overwinters as pupae in the soil. Cultivations drastically reduce numbers through desiccation. If the stubble is retained into the spring (eg as rotational set-aside or an undersown crop) then the sawflies emerge in the spring and the next generation of larvae provide food for birds.

3.14.2 Practical management

In general, management for wildlife is best undertaken using barley stubbles, rather than wheat stubbles. Stubbles of broad-leaved crops, such as oilseed rape, will also provide seed food for many species, but are less useful for yellowhammers and corn buntings, which predominantly eat cereal grains.

If the stubble holds plenty of grain and seedheads of weeds then the best

A Gouldstone (The RSPB)

Plate 3.19 The value of an overwinter stubble for birds largely depends on the abundance of broad-leaved weeds producing seeds within it.

management for wildlife is generally to do nothing for as long as possible. For stubbles preceding a spring crop, it is normally possible to leave the stubble until at least mid-February. If left unsprayed and uncultivated, birds often continue to forage in stubbles, such as rotational set-aside fields, throughout the winter and following summer.

The quality of the stubble for arable plants and birds will depend on the weed control in the preceding crop. If some form of low-input management is adopted, even if just in the headland (see section 3.12 on conservation headlands or 3.13 on low-input crops), then the broad-leaved weed flora is likely to be more diverse and abundant, and the quality of the foraging habitat improved for birds. Pre-harvest glyphosate treatments generally remove most of the plant interest and reduce the food available to birds. The decision as to whether weedy stubbles can be tolerated will generally be made on the contents of the seed bank. If it consists largely of grass weeds and cleavers, then few wildlife benefits will be gained and the agronomic consequences will be severe. If it consists of less competitive broad-leaved weeds then the wildlife benefits will be much greater and the agronomic consequences more manageable. Most scarce arable plants are associated with light soil types.

One option is to undertake a light scarification using discs or tines after harvest to unearth seeds in the soil and generate germination of arable plants that are allowed to grow through the winter. The downsides of this are that some weeds from the preceding crop will be lost, and weed seeds on the surface will be buried or germinate, either way becoming inaccessible to birds. This may be an option worth trying if the stubble does not contain much seed food and is little used by birds. If undertaken, it is probably best to scarify the field sequentially so that new seeds are made available through the winter.

The stubbles of undersown crops are less beneficial for seed-eating birds because the grass crop will limit the accessibility of the seed food for the birds (although the insect abundance is high in the spring, providing food for chicks). Stubble turnips are a useful crop for birds, particularly if little or no weed control is undertaken, as birds get access to the weed seeds as the crop is grazed off (see section 3.18).

Grazing of the natural regeneration on a stubble field may remove some of the weeds that will provide seed food later in the winter. However, if the regeneration becomes tall and thick then birds may stop foraging in it as access to the seed on the ground is inhibited and vigilance against predators is impeded. If this occurs, then light grazing may benefit birds by keeping parts of the sward open. In terms of wildlife benefits, grazing should only be considered if the regeneration is tall and dense, and does not consist of important seed-bearing plants.

Stubble height may affect its use by birds. Tall stubble (over 12 cm) may be preferred by species seeking cover from predators, such as gamebirds and skylarks, whilst shorter stubble may benefit species that depend on vigilance and flying into hedgerow cover when predators approach, such as sparrows, finches and buntings. A range of stubble height around the farm will benefit the widest range of species.

3.15 Set-aside

Rotational set-aside can provide habitat for seed-eating birds, ground-nesting birds and scarce arable plants, depending on the intensity of weed control in the preceding crop and how late that management can be delayed.

Non-rotational set-aside can be used to buffer important habitats, boost insect abundance and provide habitat for ground-nesting birds and mammals. The important principles are to establish native vegetation and minimise the amount of management undertaken during the nesting season.

Derogations can be sought to establish nesting plots for lapwings or stone-curlews, germination plots for scarce arable plants or fertile grassland for grazing winter wildfowl. Wild bird cover, woodland and short-rotation coppice can also be created on set-aside and may provide wildlife benefits if located in the right place and managed in the right way.

3.15.1 Wildlife use

Rotational set-aside

Rotational set-aside can provide significant benefits for wildlife. From harvest of the preceding crop and through the winter, the natural regeneration can provide seed food for birds. The winter benefits are dependent on the stubble containing sufficient seed food and broad-leaved weeds.

From April onwards, the natural regeneration provides nesting cover for skylarks, and potentially for other species such as lapwings, if the sward is short enough (about 10 cm) or corn buntings if tall enough (over 30 cm). The period without cultivation enables sawflies to complete their life cycle, as they overwinter as pupae in the soil and emerge as adults in about May. The adults then lay eggs and the larvae, which feed on the natural regeneration (if it has not already been removed), are important food items for birds, particularly grey partridge chicks. The summer benefits of rotational set-aside are dependent on the timing of management.

From harvest through until the green cover is removed, set-aside will provide good foraging habitat for a range of birds and mammals, including the brown hare and Irish hare, which prefer the diversity of plants and sward structure found in set-aside compared to cropped fields. It is potentially useful for the germination and seeding of arable plants, which are in turn food plants of a range of insects.

Non-rotational set-aside

Green cover on non-rotational set-aside can also be created by natural regeneration, in which case it may offer the same benefits as rotational set-aside in the first year, but develops a perennial species sward over time. Alternatively, it can be established by sowing a grass mix, which can provide a range of benefits, depending on the composition. The long-term nature of the vegetation means that non-rotational set-aside can build up abundant invertebrate populations, akin to grassland. The lack of urgency to manage the land for a subsequent crop means that the sward can provide safer habitat for ground-nesting birds through the breeding season.

Non-rotational set-aside is often used as field margin strips or field corners to take unproductive areas out of production or to buffer habitats, particularly water courses. These areas, like arable field margins (see section 3.9) can provide overwintering habitat for insects, nesting habitat for a range of birds and mammals and hunting habitat for birds of prey and owls.

Set-aside strips can also be used to buffer habitats such as Sites of Special Scientific Interest, watercourses, ponds, ancient hedgerows, hedgerow trees or unimproved grasslands from the effects of arable farming (cultivation, soil erosion, runoff, and pesticide and fertiliser drift).

Other options for management of set-aside to benefit farmland wildlife include:

- wild bird cover option – a mix of seed-bearing crops are sown to provide seed food for birds through the winter
- grassland option for wintering swans and geese – a derogation is sought to establish a fertilised grass sward to attract foraging wildfowl away from crops
- summer fallow – a derogation is sought to cultivate a plot for nesting lapwings or to enable germination of rare arable plants.

Alternatively, woodland can be established on set-aside using the Woodland Grant Scheme to benefit wildlife associated with woodland, rather than farmland. Also, short-rotation coppice can be established for biofuel. This can also benefit some species associated with wooded habitats notably some insectivorous birds.

3.15.2 Practical management

The latest explanatory guide on set-aside management should be checked in conjunction with the following advice when planning set-aside management for wildlife to ensure compliance with current set-aside rules.

When selecting land to put into set-aside, especially if adopting non-rotational set-aside, consider areas adjacent to wildlife habitats such as wetland, watercourses, woodland, mature hedgerows or unimproved grassland. The set-aside will act as an effective buffer zone and, in the case of unimproved grasslands, may even allow colonisation of plants from the grassland into the set-aside to produce a species-rich sward.

Rotational set-aside for winter seed food and summer nesting and insect foraging habitat

Rotational set-aside is best managed as a whole-field option as many species will make use of the food it provides close to the field boundary, whereas skylarks and corn buntings are more likely to nest and feed in the centre of the field.

Generally, natural regeneration will provide food for birds and mammals from harvest of the preceding crop until the time when the green cover is removed, either by spraying or cultivating. The amount of seed food for birds will depend on the amount of grain spilt during the preceding harvest and the weediness of the cover, which is largely dependent on the intensity of weed control in the preceding crop. The abundance of broad-leaved weeds is a critical factor. Grass weeds (other than annual meadow grass) and volunteer cereals provide very little food value for birds, mammals and insects, although volunteer cereals can provide important cover for grey partridges. If broad-leaved weeds can be encouraged in the preceding crop by adopting some form of low-input management or conservation headlands, then the quality of the natural regeneration for wildlife will be improved. Pre-harvest glyphosate or any post-harvest herbicide application will reduce the wildlife value.

Winter grazing will keep the sward short. This is beneficial in terms of giving birds access to the ground where they forage for seed food, but the grazing may remove the seed-bearing weeds. In general, it may be advantageous in situations where volunteer cereals and grass weeds create a tall, dense cover. Autumn cultivations to create a false seed bed will encourage the germination of autumn-germinating weeds, but will also remove some of the weeds from the previous crop and bury seed food on the soil surface. For maximum wildlife benefit, it would be ideal to lightly cultivate up to half of the area, but leave the remainder to provide seed food in the early winter.

Set-aside rules allow the removal of the green cover by spraying, cutting or cultivating (set-aside rules stipulate dates when each form of management is allowed). Spraying and cultivating generally remove arable plants and reduce the available food plants for insects and seed and insect food for birds and mammals. Cutting and cultivating are likely to destroy the nests of birds, and are generally not permitted until late in the nesting season. Where possible, avoid cutting rotational set-aside to avoid nest destruction. Spraying does not destroy nests directly, but the nesting cover and the food availability decline over the subsequent two weeks. This may result in increased nest failure rates, and will prevent subsequent nesting attempts.

Spraying off the green cover is often undertaken at the earliest opportunity in the spring. This will minimise the potential for ground-nesting birds to use the set-aside, with the possible exception of lapwings (although plots cultivated in February are more likely to attract them – see below). If the vegetation is largely composed of broad-leaved weeds and volunteer cereals, then the need to control the weeds is less urgent and can be delayed until later in the breeding season. If grass weeds are abundant, then the optimum time to spray them off is at the start of seeding. In southern England, this is likely to be the end of May for black grass and barren brome, early June for wild oats, and July for couch grass. Timings will be later further north. If it is possible to leave some headlands unsprayed until later in the year, then this will benefit birds making use of the field margins.

The timing of cultivation of rotational set-aside is critical for ground-nesting birds in the field. The later, the better, as skylarks and corn buntings can still have chicks in the nest in August. If early weed control or cultivation is likely to be necessary, then creating areas of non-rotational set-aside as ideal nesting habitat for ground-nesting birds (see below) in addition to retaining areas of rotational set-aside for winter foraging may be the best approach.

Non-rotational set-aside for summer nesting and insect foraging habitat

Non-rotational set-aside is normally sited to take out unproductive areas of the farm. This is often an unproductive field margin or field corner, but may be used as a whole-field option, in which case, it has the added benefit of creating habitat for the in-field nesting bird species, such as skylarks and, potentially, corn buntings or lapwings. Grass strips can also be created through the middle of fields to break up large arable blocks in a similar way to beetle banks – these will benefit in-field insect populations and may be used by ground-nesting birds. The current rules state that minimum width of mid-field set-aside strips is 20 m, but this may be reduced to 6 m along field boundaries bordered by a hedge or watercourse.

Natural regeneration is a better approach to establishing a sward in that it will develop native vegetation of local provenance. However, there is a risk that this will include noxious weeds such as grass weeds, cleavers, thistles, docks or ragwort. A safer option is to sow a grass mix. For maximum wildlife benefit, a wide mix of native grass species will be better than a more simple mix or use of non-native or agriculturally improved varieties. Field margin strips can be sown with a mix of tussocky and fine grass species (see section on arable field margins, 3.9, for details) Inclusion of wild flowers will provide added benefits for pollinating insects, but will also add significantly to the cost of the mix. Again, it is important to use native species. Scabious, knapweed, yarrow and ox-eye daisy are species able to survive within a grass sward, but it is better to avoid inclusion of tussocky grass species if flowering plants are to be included.

For whole-field set-aside, a mix of fine grasses, such as fescues, bents, meadow grasses and crested dogstail would be more appropriate to produce a more open sward suitable for nesting birds. A small component of tussocky grass species (maximum of 5% of seed mix by weight made up of cocksfoot, timothy or Yorkshire fog) will give a more varied sward without compromising the open grass structure for nesting birds.

Non-rotational set-aside has to be cut once a year. Currently, this is between 15 July and 15 August. It is important to follow these guidelines:

- Cut as late as possible, to reduce the number of nesting attempts that will be disrupted, as skylarks and corn buntings can still be nesting in August.

R Winspear (The RSPB)

Plate 3.20
The 25% of set-aside left uncut in this field harbours a large population of small mammals and helps support the barn owls nesting in the adjacent nestbox.

- Make use of the rule allowing 25% of each set-aside area to remain uncut (Plate 3.20). Although the same area can remain uncut for up to three years, it is better to leave a different area uncut each year so that the sward does not become too dense and rank.
- Make use of the rule allowing up to 2 m next to a hedge or wood to be left permanently uncut. This will allow a rough, tussocky margin to develop and large insect populations to build up.

Areas that are left uncut for a year or more will often hold large densities of small mammals and provide attractive hunting areas for species such as barn owls and kestrels.

Control of perennial weeds (thistles, docks or ragwort) should be done by spot treatment or using a weed wiper.

Grazing is permitted between 1 September and 14 January. Grazing can create a more varied sward structure and enhance the abundance of invertebrates within the field.

Managing set-aside for nesting lapwings or stone-curlews

Where lapwings or stone-curlews nest on or within 2 km of the farm, cultivation of fields or plots (at least 2 ha) can supplement existing nesting habitat, and ensure that there is still nesting habitat later in the spring when other opportunities, such as spring crops, have become too tall and dense. Cultivation of the green cover will require a derogation from normal set-aside rules.

If possible, use a field where the birds have nested in the past. It is advisable to locate plots on exact sites of previous nests if these sites are known. If previous nest sites are not known then select a large field, ideally next to a permanent pasture field that will be grazed through the spring and summer. Avoid siting plots close to tall hedgerows or woodland.

The plot should be cultivated annually: either ploughed in the autumn and allowed to weather down through the winter, or ploughed and cultivated with tines or discs in February or early March. If the regeneration of weeds creates a tall, dense sward then this should be sprayed off or re-cultivated (if no nesting birds are present) to maintain the condition for nesting lapwings and stone-curlews. Lush growth may attract nesting skylarks. It may also attract seed-eating birds if there is a high broad-leaved weed content, particularly turtle doves through the summer.

Managing set-aside for rare arable plants and seed-eating birds

Annual cultivation of sites with rare arable plants will promote germination and can aid conservation of the species if they can be maintained without the build-up of competition from other weeds. Cultivation of the green cover will require a derogation from normal set-aside rules. See section 3.11 on cultivated margins for best management practice.

Any cultivations that create an area rich in broad-leaved weeds will attract seed-eating birds as soon as seed food becomes available. This type of management can be used to support seed-eating bird populations, notably turtle doves that require seed food from May until September, and the suite of resident seed eaters that require seed food throughout the winter.

The wild bird cover option

A wild bird cover can be created on set-aside (see section 3.7 on wild bird cover for the best management practice), but ensure that set-aside rules are covered. Currently, these include restrictions on what can be sown, use of pesticides and limits on fertiliser use.

Managing set-aside for wintering swans and geese

An improved grass sward can attract wintering swans and geese away from cereals and other crops. They will select swards that receive fertiliser applications or include clover in the mix, and so require a derogation from normal set-aside rules.

Select a field that has been used by swans or geese in the past and that is within 10 km of a known roost. The field should be at least 10 ha. A total of 15–30 ha are required per 1,000 birds Two or three separate areas are better than a single block as the birds can move between sites if disturbed. The recommended mix is perennial ryegrass, timothy and white clover at a ratio of 5:1:1 with 5% of fescues at a total seed rate of 30 kg per hectare. Graze or cut the sward to a height of 5 cm in September and remove the cuttings if possible.

3.16 Fallow plots for ground-nesting birds

Summer fallow plots for nesting lapwings or stone-curlews are best sited where birds have nested before. They should be left as rough fallows from mid-March until at least mid-July (later for stone-curlews). Summer fallows may also benefit scarce arable plants and seed-eating birds, although these may benefit equally from cultivated margins.

3.16.1 Wildlife use

Cultivated, uncropped summer fallow plots provide ideal nesting habitat for lapwings and stone-curlews, and they should be specifically targeted at sites where these species are likely to be attracted (Plate 3.21). The birds may also use them for foraging for invertebrates through the breeding season if there is no grazed pasture nearby. The plots may also provide suitable conditions for rare arable plants, although margins are likely to be better locations for these (see section 3.11 on cultivated margins). The natural regeneration on the plots is also likely to attract seed-eating birds, especially if it can be left fallow through the winter and cultivated in February or early March. Turtle doves are likely to feed on any weed seeds produced through the summer.

3.16.2 Practical management

Summer fallow plots are best sited in the exact locations where nesting lapwings or stone-curlews have been recorded before. If they nest in the area, but the exact nest sites are not known, then plots are ideally placed in arable fields adjacent to grazed pasture in fields of at least 5 ha without tall hedges or woodland boundaries. For stone-curlews in particular, proximity to busy roads reduces the chances of their uptake. Avoid them within 1 km of a trunk road unless recent nesting at this location has been recorded.

The plots should be at least 2 ha in size and at least 100 m wide. Do not cultivate fallow plots on grassland with existing wildlife interest (eg unimproved or species-rich grassland). However, plots within grazed pasture fields would potentially provide an ideal combination of nesting and feeding habitat. Fallow plots on grassland would need fencing if livestock will be grazing the field during the nesting season.

David White/rspb-images.com

Plate 3.21 Lapwing brooding chicks on a summer fallow plot.

Summer fallow plots can be in fixed locations or rotated around suitable sites on the farm. If a successful site is selected, then a fixed plot is a better option as birds are more likely to return to it year after year. If it is necessary to create a plot in a rotation, then a two-year rotation between the two halves of a field of about 10 ha can work well. A fixed plot has the agronomic advantage of not returning a potentially weedy plot to arable cropping. There is also a risk of wheat bulb fly in a cereal crop following a summer fallow.

The plot should be cultivated annually. It should either be ploughed in the autumn and allowed to weather down through the winter, or ploughed and cultivated with discs or tines in February or early March to produce a rough, bare surface. The advantage of spring cultivations is that the preceding stubble or regeneration on the previous year's fallow can provide important winter food for a range of seed-eating birds, but depends on the land being workable before mid-March. The plot should not be rolled, or worked down to a fine tilth. The ideal surface is even enough to allow lapwings to move around freely and be able to see all around them, but not too flat so that sitting birds are clearly visible to predators (Plate 3.22).

The plots should remain undisturbed between mid-March and mid-July. Unlike lapwings, stone-curlews can have second broods, and so plots managed for them should be left until September. If regeneration is dense and exceeds 10 cm in height early in the season then lapwings are unlikely to use the plot. It may still have value for arable plants or seed-eating birds. However, if the primary objective is nesting lapwings and no other wildlife interest has developed then it may be useful to spray off the regeneration with glyphosate or re-cultivate the plot (if confident that there are no nesting birds present). It may be necessary to spray off the natural regeneration prior to cultivating the plot in subsequent years.

R Sheldon (The RSPB)

Plate 3.22
The surface should be rough enough to provide nesting birds with some cover to conceal themselves.

3.17 Skylark plots

Skylark plots are undrilled patches of 16–24 m^2 created at a rate of two per hectare through winter cereal fields to boost skylark breeding success. Cereals drilled at a conventional seed rate but with double spaced rows (25 cm apart) were also tried as a way of opening up the canopy for foraging skylarks, but the results showed less consistent benefits.

3.17.1 Wildlife use

The undrilled patches in a winter cereal crop are designed to open up the canopy sufficiently to provide regular foraging sites for skylarks when they are feeding young. In a conventional winter cereal field, the crop has become so tall and dense by June that more than half of the birds' foraging time is spent outside of the field. If adjacent fields are also winter crops, then birds will struggle to find sufficient food. In fields with two undrilled patches per hectare, skylarks continue to feed within the field throughout the breeding season.

These patches also have higher densities of some insect groups that birds feed on, but the main benefit is the sparse vegetation that enables easier access to the insect food. The research results indicate that fields with two plots per hectare had a greater density of nesting skylarks than conventional fields, birds made nesting attempts later into the year and the nests were also generally more productive in fields with the plots. Overall, skylark productivity was increased by almost 50% in fields with two plots per hectare (taking up less than 0.5% of the field area) (Plate 3.23). The skylark chicks in fields with plots were also heavier, suggesting that they were fitter and more likely to survive through the winter.

Given the increased abundance and accessibility of insects in the plots, there are also potential benefits for other birds that feed on insects within the crop, such as corn buntings and yellow wagtails.

Andy Hay/rspb-images.com

Plate 3.23
Two plots per hectare, each of 20 m^3, represent 0.4% of the field area, but can boost skylark productivity within the field by almost 50%.

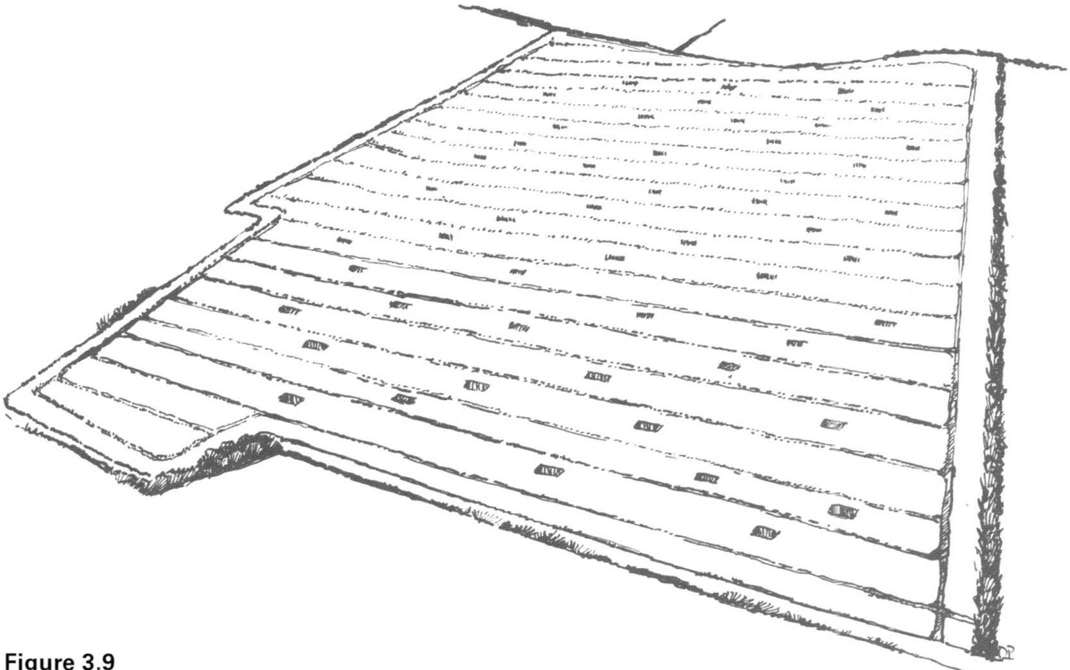

Figure 3.9
A 25 ha winter cereal field should ideally contain about 50 plots scattered across it. The plots should not be within 24 m of the field edge or connected to the tramlines.

3.17.2 Practical management

This feature is designed for use in winter cereal fields only. Skylarks generally avoid small fields. The best fields to use are ones of at least 5 ha if bounded by open field boundaries or short hedges, or fields of at least 10 ha if bounded by tall hedges or woodland.

The plots should be put in at a rate of two per hectare. They should be located at least 25 m from the nearest edge (not in the headland), not connected to the tramlines, and away from features such as telegraph poles (Figure 3.9). The plot size should be between 16 and 24 m^2. See Table 3.12 for details of the ideal plot length for different drill widths.

When the seed drill is switched off or lifted, there may be some seed spillage causing a gradual transition to bare ground, but this is to be expected and does not compromise the effectiveness of the plot.

After drilling, the skylark plots are generally managed in the same way as the rest of the field. They receive the same pesticides and fertilisers as the rest of the field. If difficult weeds such as black grass grow within the plots then these can be sprayed out with a knapsack sprayer, but this was rarely necessary in the trials.

Table 3.12 Ideal plot length for different drill widths.

Drill width (m)	Plot length (m)
3	6–8
4	4–6
6	3–4
8	3

3.18 Arable pockets in pastoral areas

Cereal and fodder crops can provide significant benefits to birds in pastoral areas, particularly seed-eating birds. The wildlife benefits of maize, however, are not as great as those of some other crops. Where arable crops cannot be incorporated into the farming system, small plots of wild bird cover can be very beneficial.

3.18.1 Wildlife use

Growing cereals or fodder crops in livestock systems provides an element of habitat diversity important for some birds and other wildlife (Plate 3.24). In particular, they are a source of seed food for birds such as the grey partridge, skylark, tree sparrow, corn bunting and yellowhammer. These species have declined considerably in pastoral areas as seed food availability has dropped with the loss of arable cropping and intensification of grassland management. Ripening grain is a valuable chick food for buntings and sparrows in cold, wet weather, when invertebrate prey is not available.

In addition to seed provision, cereals or whole-crop silage have other benefits for birds. Cereals, particularly spring crops, provide potential nesting sites for a number of species. The lapwing is attracted to bare ground or short, open vegetation. Winter-sown cereals tend to be avoided. Cereals also provide good nesting habitat for skylarks. They are generally more successful in spring-sown cereals, as the crop stays shorter and more open for longer into the season. As agricultural grasslands have become less suitable nesting habitats for these species, arable crops have become increasingly important.

Undersown cereals provide similar benefits to those stated above. When sown in spring, they can be attractive to breeding skylarks and lapwings. Cereal grains provide important seed food, while invertebrate prey can be abundant before and after harvest. The inclusion of clover is likely to be particularly beneficial. Many insects feed on clover or take nectar from the flowers. Clovers have more chance to flower and seed when undersown, compared to a grazed or mown clover-grass sward. The spring sward following undersown cereals can be particularly rich in invertebrates such as moth and sawfly larvae. Winter stubbles of undersown cereals provide far more seed food than straight grass reseeds, but dense vegetation may make seeds less accessible than cereal stubbles which are not undersown.

M Hancock (The RSPB)

Plate 3.24 Arable crops within a grass-dominated landscape provide habitat for annual plants and seed food for birds.

Fodder brassica and root crops such as stubble turnips, fodder beet, rape, kale and swedes can be particularly beneficial in providing winter food for seed-eating birds. It is the broad-leaved weeds within the crops, rather than the crops themselves, that provide the seed food. In addition, thrushes, pipits and wagtails favour brassica cover later in the winter. Invertebrate prey are more numerous and accessible in the relatively warm, wet microclimate of such crops. Fodder brassica and root crops are also used for roosting and cover. Brassica crops may also be used as a source of insect prey by nesting birds such as yellowhammers. Stubbles following fodder brassica and root crops can also provide nesting habitat for lapwings if cultivation of the stubble is sufficiently late.

3.18.2 Practical management

Areas where arable cropping is not appropriate include fields susceptible to soil erosion, archaeological sites, unimproved permanent grassland and grassland with existing wildlife interest.

As the amount of cultivation taken up on livestock farms is likely to be limited, it is important that if arable pockets are created, they provide as many conservation benefits as possible. Spring-sown cereals, managed with minimal inputs, and the stubble retained over the winter have the potential to provide the widest range of benefits for birds. Ideally, to attract lapwing, sowing should be completed by mid-March, though later crops can still attract birds to re-nest when they have failed elsewhere. Fields in open areas away from woodland are more attractive to nesting lapwings and skylarks. However, a wide range of fodder crops can provide important resources for birds if managed sensitively. Funding for some of these crops may be available from agri-environment schemes.

Cereal-based whole-crop silage

Cereals for whole-crop silage can be drilled in spring or autumn. Wheat and barley are the most popular, but oats, rye and triticale can also be used. Ensiling cereals for whole-crop silage has increased in popularity as preservation and harvesting techniques have improved. Benefits over other forage options include little or no silage effluent, high feed value and lower costs of production associated with a single harvest per year. In addition, cereals can be grown in most parts of the country, and there is flexibility with the choice of ensiling or combining at harvesting. Cereals can also be grown with pulses and brassicas as a bi-crop to raise the protein content of the silage.

The harvesting of whole crop silage before grain has fully ripened means that little grain is spilt at harvest. The early harvest may particularly affect corn buntings, which nest late in the summer in cereal crops. Combined spring cereals are more beneficial for them. Alternatively, a cereal-based wild bird cover crop can be grown.

Maintaining stubbles over the winter can provide vital seed sources that aid winter survival for a number of seed-eating birds. Grain within silage fed around stockyards or to out-wintered stock can also be a valuable source of food through the winter (Plate 3.25).

All cereal-based silages provide important nesting and foraging habitat on pastoral farms, offering many more benefits to

R Winspear (The RSPB)

Plate 3.25
Cereals fed to out-wintered stock also provide a feeding opportunity for seed-eating birds, particularly in late winter and spring, when food on stubbles has become depleted.

wildlife than grass and maize silage. However, applying no insecticides to spring crops after 15 March, and limited broad-leaved weed control, would maximise the benefits for wildlife (see section 3.13).

Undersown cereals

Undersowing cereals with grass or a grass/clover mix used to be a widespread practice in agriculture, providing a simple and cost effective re-entry into grass from arable crops. With the loss of livestock from arable farms and more reliance on silage in the livestock sector, its use has declined rapidly. However, developments in whole-crop silage and crimping of grain provide different harvesting methods that do not rely on grain ripening. Undersown cereals offer an alternative high quality feed. They offer considerably more wildlife value than directly reseeded grassland.

Forage maize

Maize has become a major crop in UK agriculture, with the development of hardier and earlier maturing varieties. While maize does provide some habitat diversity in grass-dominated areas, its benefits to wildlife are not as great as those of some other crops.

The vulnerability of maize seedlings to weed competition means that the crop requires comprehensive weed control. This results in very few seed and insect resources being present within the growing crop. Consequently, stubbles after harvesting are usually relatively clean, offering little food for seed eaters. However, there may be some maize grain and surface seeds immediately after harvest to attract seed-eating birds for a short period. Grain within maize silage is taken by seed eaters such as tree sparrows in winter.

Lapwings looking for prospective nesting sites in March find maize stubble and newly cultivated fields attractive nesting sites. However, their breeding success on maize crops can be very low without careful management. Lapwing chicks hatch about a month after the birds start incubating. Often, they are still sitting when cultivations for sowing begin in April/May, resulting in many nests being lost. Spreading slurry when birds are sitting can also cause nest loss. Pairs can re-lay successfully in the new crop, but cultivation and sowing operations are often protracted. This can destroy the first clutch of eggs and the following replacement clutch. Field operations in the spring should be concentrated into as short a time as possible, ideally within a week. Cultivating around nests can make them more obvious to predators. Further west and north in the UK, cultivations may start late enough for chicks to hatch before the onset of cultivations.

For birds nesting in crops that have been drilled, hatching success can be high. However, maize fields are often poor foraging habitats and chick mortality due to starvation is sometimes high. Habitat such as damp pasture or shallow-edged ditches at the field boundary or in accessible adjacent fields will help chick survival. The structure of maize makes it unsuitable for most other ground-nesting birds.

Fodder brassica and root crops

Brassica and root crops are still important fodder crops in some areas, and the weeds within them provide a good food source for seed-eating birds (Plate 3.26).

R Winspear (The RSPB)

Plate 3.26
Brassica crops grown with little or no herbicide treatments harbour a wealth of broad-leaved weeds that provide seed food for birds, particularly as the brassica crop is grazed.

While crops can be susceptible to weed competition during establishment, a low to moderate weed infestation later in the year has little affect on yield. Crops drilled directly into a desiccated grass sward contain fewer weeds. However, as well as reducing their value to seed-eating birds, such crops tend to be lower yielding than those sown into cultivated seed beds. Weed seeds are taken from within the crop, but become more accessible after the crop is grazed. Strip grazing regularly opens up new feeding areas through the winter.

An abundant weed flora within these crops can also boost the invertebrate abundance, which benefits the insectivorous birds that forage within them.

Other fodder crops

Other fodder crops such as red clover and lucerne can also be valuable habitats for birds. Grey partridges eat clover seed, and the pollen and nectar produced by flowering clover attracts insects. Where arable crops cannot be incorporated into the farming system, small plots of wild bird cover would be an alternative way of providing seed food.

Further reading

Hancock, M H and Wilson, J D (2003) Winter habitat associations of seed-eating passerines on Scottish farmland. *Bird Study* 50: 116-130.

3.19 Grassland management for birds

Management that increases the abundance of invertebrates and seeds in grassland will be particularly beneficial to birds. Grazed pasture is favoured by birds that require short swards for nesting or feeding, but many have strong preferences for certain sward features. Nesting birds that require tall cover prefer mown meadows, but the timing of cutting is critical.

Avoid management for particular species if the situation is inappropriate for them. Grasslands with special wildlife value have often developed because of historical management, so consider the implications for existing wildlife before changing management regimes.

3.19.1 Wildlife use

Lowland grasslands are used for nesting and feeding by a range of birds. They are also used for feeding by a number of resident species, such as the chough, song thrush and starling, and winter visitors, such as the golden plover, redwing and fieldfare. Many more species rely on the grassland element of mixed farming landscapes to provide a rich source of invertebrate food (Plate 3.27). Birds have strong preferences for certain types of swards, depending on their feeding and nesting requirements (Table 3.13).
The location of the grassland is critical when considering the requirements of some birds, particularly nesting waders, which require relatively large, open fields away from tall features, such as woodland or tall hedgerows.

Grazed pasture is favoured by low-growing plants, insects associated with animal dung and birds that require short swards for nesting or feeding. Invertebrates associated with dung degradation such as flies and beetles are important food for waders (eg lapwing and curlew), swifts, swallows, wagtails, starlings, crows (eg chough and rook) and many species of bats. Earthworm densities, particularly important for waders, are also higher on grassland with inputs of animal dung.

Mown meadows are more likely to produce flowers and seeds, and are preferred by nesting birds that require tall cover.

Unimproved grasslands hold a much greater diversity of wildlife than agriculturally improved grasslands. The

Plate 3.27 Large numbers of birds come to feed on grasslands rich in invertebrate food.

Table 3.13 Sward structure requirements of some grassland birds.

Species	Nesting	Summer foraging	Winter foraging
Whinchat	Very lightly grazed to ungrazed vegetation. Sometimes amongst sparse bracken and in association with sparse scrub		
Curlew	Plenty of cover, eg damp, unimproved hay meadows, upland rough pastures		
Snipe	Wet pastures and peat bogs with a mosaic of taller vegetation (>30 cm) for nesting, interspersed with short vegetation (<10 cm) for feeding. Soil soft enough to penetrate easily with a 6-inch nail		
Redshank	Mosaic of short, damp grassland, 5–15 cm high, with some taller tussocks. Some standing water through the breeding season, particularly in May and June		
Skylark	Large, open fields with sparse vegetation around 20–50 cm in height, eg lightly grazed pastures or hay meadows		Large, open fields with plants containing seed
Meadow pipit	Swards of 10–50 cm, possibly including sparse, open scrub		Moderately tall, gappy, grazed swards
Lapwing	Short grassland (average sward height around 5 cm) with occasional scattered taller tussocks (<10% of area). Chicks are often moved to feed at damper flushes and edges of temporary or permanent pools		Short, open vegetation less than 5 cm and bare earth providing access to soil invertebrates, eg permanent grassland grazed by cattle
Yellowhammer		Structurally diverse swards containing longer-lived tussocks, plus patchy short vegetation (<3 cm) and bare earth	Fields with plants containing seed
Song thrush		Short, open vegetation less than 5 cm and bare earth providing access to soil invertebrates near large hedgerows/woodland. Particularly cattle-grazed pasture and wet areas	
Starling		Short, grazed vegetation less than 5 cm tall providing access to soil invertebrates, eg permanent grassland. Also attracted to swards containing plants in seed	
Linnet		Swards with flowers producing seed (especially dandelions, sorrel)	Fields with seeding broad-leaved plants
Wheatear		Tightly grazed swards 1–3 cm in height	
Stone-curlew	Tightly grazed lowland dry grassland (<3 cm) with sparse vegetation and bare areas		

diversity of wildlife in grassland is also boosted by the range of features, such as variable sward heights, tussocks, presence of bare ground, and variation in topography and water levels.

3.19.2 Practical management

Grazing

Different birds prefer different sward structures to feed and nest in, depending on how they feed and how they avoid predators. Some birds have other requirements for their breeding habitat, such as a certain level of soil wetness or larger open fields. Targeting the right sward structure to appropriate fields will be important for these species.

Two key resources that have become particularly scarce in agricultural grasslands are seeds and large, long-lived invertebrates, as few plants are free from cutting or grazing for long enough to seed or harbour long-lived insects throughout their life cycle. The most valuable pastures contain a variety of sward structures. Short grass provides easy access to food resources and open terrain to enable vigilance against predators, while tufts of grass and longer-lived tussocks provide cover, harbour more insects and may produce seeds (Plate 3.28).

Extensive seasonal grazing provides stable conditions suitable for many invertebrates, but rotational grazing can allow certain areas to be avoided at sensitive times.

Areas where animals congregate such as sheep tracks, gateways and drinking points provide short vegetation or bare areas that offer niches for some plants and invertebrates. Many birds are attracted to such bare areas where they find food, such as seeds of annual meadow grass and accessible insects.

Dung from livestock adds to the invertebrate interest of a grazed field. The use of anthelmintics is a necessary part of livestock farming, with different families of anthelmintic needing to be rotated to avoid the build-up of resistance. Any management that helps to reduce the use of these chemicals will be beneficial to wildlife, as well as saving the farming business time and money. As most parasitic worms are host-specific, mixed grazing systems can break the life cycle of the parasite. In addition, taking faecal egg counts, avoiding overstocking areas and the judicious use of clean grazing all help to avoid unnecessary treatment. Of concern, through their residual properties, are the effects of the avermectin family on dung-degrading invertebrates and the potential knock-on effects to wildlife that feed on these. Studies are in progress to determine the true effect. Moxidectin is less harmful than some other members of the avermectin family. Where possible, try to avoid treating stock at pasture in the spring. It may be possible to treat cattle at housing. Other anthelmintic families do not appear to have an adverse effect on dung fauna.

Different livestock types create different sward structures and plant composition.

- Sheep are quite selective grazers, favouring the most nutritious and palatable vegetation in a sward. At low densities, they produce a mosaic of closely cropped areas interspersed with areas of ranker vegetation. At higher densities, grasses produce more tillers and lower growing plants are favoured at the expense of plants that cannot tolerate heavy grazing pressure. This results in very short, dense swards. Sheep also favour flower heads, reducing nectar sources and seed production.
- Cattle are less selective grazers, eating a wider range of vegetation. This results in grazing pressure being more evenly spread across a grazing area. They tear

D Buckingham (The RSPB)

Plate 3.28 Seeds and invertebrates will be most abundant in structurally diverse swards.

vegetation rather than biting it, creating a more open, less densely tillered sward. This, coupled with the heavy trampling, is more likely to expose soil, offering opportunities for new growth and germination. Such areas are also favoured for foraging by several birds. Fields grazed by cattle support higher densities of soil invertebrate-feeders like thrushes. A natural tendency for cattle to avoid grazing near cowpats to reduce parasite burdens creates a more tussocky sward.

- Horses graze some areas very tightly while completely leaving others. Dunging in the same area leads to local nutrient enrichment. This affects the composition of the vegetation, favouring vigorous, coarse growth. The result is poor plant diversity, but some grasshoppers favour the mix of tall and short vegetation. Horses will also eat rough vegetation, including scrub.

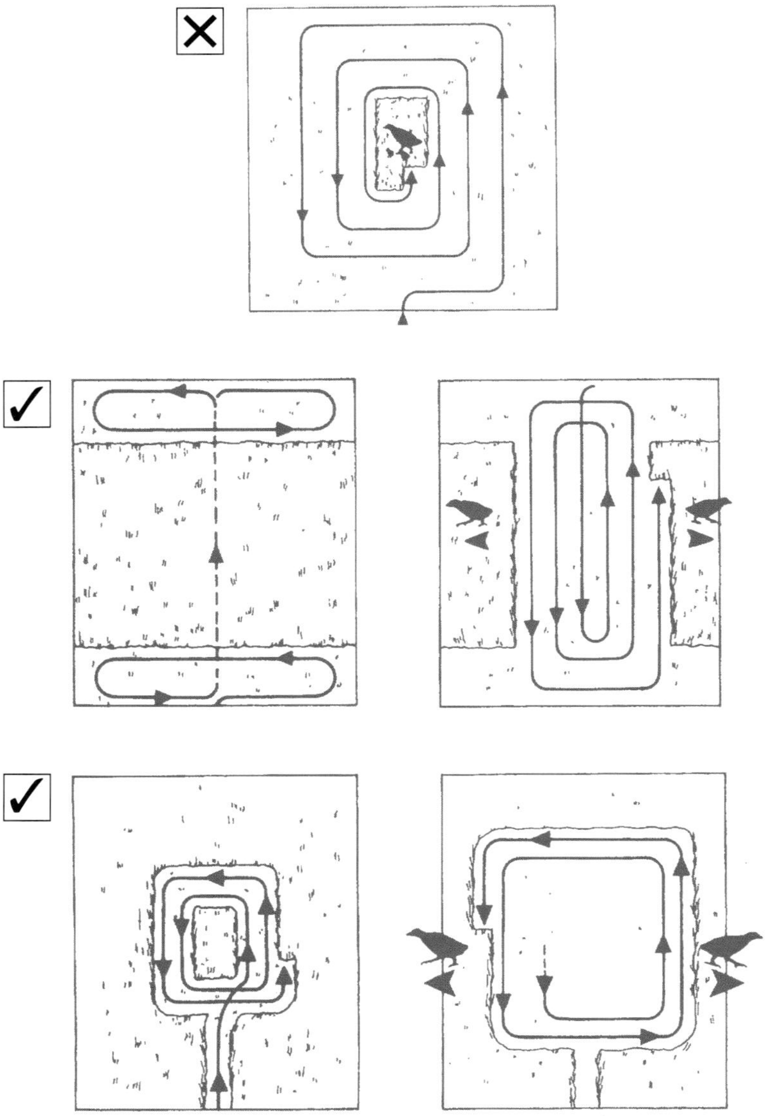

Figure 3.10
Mowing techniques to avoid destroying corncrakes, their eggs and chicks. Cutting from the outside of the field inwards isolates the chicks in the centre of the field. Cutting from the inside outwards gives chicks the chance of escaping to the field edge.

Sometimes nutritionally poor herbage needs to be grazed. It is important to match the herbage quality with the nutritional requirements of the stock, which will vary with factors such as the age and stage in the breeding cycle of the animals. Mineral supplements may also be required.

Mowing

The curlew will nest in meadows, but often cutting is carried out too early for it to complete breeding. The curlew needs cover to be maintained until mid-July. It is unrealistic to expect an agricultural system to refrain from cutting the majority of its fodder fields until this late in the season, but cutting fields known or likely to be favoured by curlew later would protect it in important nesting fields. In the lowlands, the curlew generally favours damp, unimproved meadows. Some later-cut fields would also provide important resources for other species, such as nesting habitat for skylarks and food for seed eaters.

The corncrake was once a widespread species in the British Isles, but with the onset of mechanisation and the switch from hay to silage, it died out rapidly. Where it persists, in north and west Scotland, targeted funding has enabled fields to be specifically cut late in the season and in a corncrake-friendly manner (see Figure 3.10). This has helped boost the population considerably over the last decade.

Skylarks will sometimes nest in silage fields, but early cutting destroys nests and subsequent cuts can destroy following nests if they are spaced too closely. For the first cut to avoid destroying a large proportion of first broods, cutting would need to be delayed until late May. Skylarks nest in vegetation once it reaches around 20 cm. They then require 25 days for chicks to leave the nest, plus an additional period before the young birds are able to fly. This means a requirement for at least a seven-week period between cuts.

Early cutting destroys nests and chicks, whether fields are intended for hay, haylage or silage. Late cutting of mature grass for haylage or silage has some of the wildlife benefits of a traditional hay meadow.

Mown fields generally have a lower diversity of invertebrates than grazed pasture, as only certain species can adapt to the repeated, wholesale removal of vegetation. However, mown grasslands can provide nectar, pollen and other food resources that can be visited by mobile invertebrates if left long enough to flower. Opportunities are greatest in swards containing a diversity of grasses and flowering plants.

Plants that have gone to seed in the sward before cutting provide food for seed-eating birds. Traditional hay meadows can provide an abundance of seed food. Linnets favour the seeds of broad-leaved plants, such as sorrel and dandelions. The fodder produced by seed-rich meadows can also provide birds with food through the winter when the fodder is fed to livestock, and in the resulting manure.

Fields cut for zero-grazing and silage usually provide few seeds, but uncut margins and field corners can provide some seed. Leaving uncut margins and corners in mown fields provides a refuge for many insects and can also create ideal habitat for small mammals, which are in turn hunted by barn owls and kestrels (see section 3.24). Grazed pasture can also contain populations of small mammals where swards are not grazed very tightly. It is estimated that a breeding pair of barn owls require around 1.5 ha of rough grass within a territory. Territories vary, with most hunting taking place within 2 km of the nest during the breeding season and ranging up to 3–4 km from the roost sites in the winter as prey becomes scarcer.

Mechanical operations

Mechanical operations on grassland such as muck spreading, rolling, topping and harrowing can be particularly damaging to ground-nesting birds in the spring. Wherever possible, they should be timed before or after the breeding season in fields with breeding birds (see Appendix 1 for the timing of breeding of birds in grassland). Some operations, such as topping, can be damaging to invertebrates through reducing sward patchiness or by damaging features such as anthills. Rolling reduces populations of leatherjackets and molluscs, important prey for the starling and song thrush respectively.

Improved grassland

There have been a number of changes in grassland management (see section 1.5). If bird declines are to be reversed and former ranges recovered, the resources required by these birds will need to be fitted within modern livestock systems. Small modifications to current practices in agriculturally improved grassland may be able to provide significant benefits to some species.

The variety of grassland structures provided within agricultural systems determines the variety of resources provided for wildlife throughout the year. Increasing plant diversity in productive swards enhances sward structure through the individual growth habits of the different plants. Even the inclusion of a couple of grass species other than ryegrass, such as timothy, cocksfoot and red fescue, will have benefits. Additional plant species also add new food resources to the sward, such as seeds at different times of the year or specialist insects that only feed on certain plants. Several grasses, such as Yorkshire fog, red fescue and creeping bent, are as productive as perennial ryegrass under low levels of fertiliser inputs, and the presence of legumes and herbs in the sward can extend the mineral composition of the herbage for grazing livestock.

Farmyard manure is the best fertiliser to use from a wildlife perspective, as it boosts soil invertebrate densities. Important prey items for birds such as earthworms are associated with areas where manure has been applied in the previous season. Solid farmyard manure has the positive effect of bringing soil invertebrates closer to the surface, thus making them more accessible to birds. It also has less effect on the plant diversity of the sward than slurry, which releases its nutrients quicker. Around 20 t/ha of farmyard manure every three to five years can be applied to botanically-rich hay meadows where it has been traditionally practised without adverse effects on the flora. A greater reliance on clover than fertiliser brings benefits. Where they mature and flower, clovers can provide seeds and boost the abundance of invertebrates.

Cultivations disrupt the life cycles of invertebrates. Some of the smaller, more mobile invertebrates can recolonise a sward quite quickly, while larger, longer-lived invertebrates that are of particular value to birds take several years to return, even if ongoing management is favourable. Arable cropping is beneficial to seed-eating birds, but maintaining some permanent grassland and long-term leys within the farming system will be beneficial to insectivorous birds. Partial reseeding of old grass leys, where some desired plants are drilled directly into the existing sward, will be less damaging to invertebrates than cultivating the pasture. Permanent grassland supports high densities of wintering golden plovers, redwings and fieldfares. This is probably due to the high densities of beetles and earthworms. Thrushes prefer permanent grassland surrounded by large hedgerows that offer safe cover from predators, whilst waders prefer more open locations where approaching predators can be seen from greater distances.

Seeds are scarce in modern agricultural grassland, with seed eaters in pastoral areas very dependent on the unripe grain and weed seeds in remaining arable fodder crops. However, grassland will also be used where seeding grasses and broad-leaved weeds can be found. Grey partridges, skylarks, yellowhammers and reed buntings will all feed on the seeds of many grasses, while species such as the linnet and the greenfinch are more dependent on seeds of broad-leaved plants. New leys following arable crops may be used by seed eaters, particularly undersown leys that retain some arable weeds. Fertilised ryegrass swards can produce large quantities of seeds if no mowing or grazing occurs for long enough to enable the grasses to go to seed. Initial experiments have shown that ryegrass swards allowed to mature and seed late in the growing season can provide important seed food over the winter for seed eaters such as the yellowhammer and the reed bunting. Dense grass needs to be removed in the following spring to avoid sward damage and allow fallen seed to germinate.

Grazing geese and wintering wildfowl favour improved and fertilised grasslands for the higher protein content in the forage. Their preference is for tightly grazed swards and the targeting of such grassland in their localised wintering areas will benefit these species.

Grassland of high conservation value

Grasslands with particularly significant wildlife value include areas supporting rare or localised species, seasonally damp areas or unimproved grasslands. Such areas, even if small, should be valued and managed with consideration of the wildlife interest. Avoid practices that may be damaging, such as draining, pond excavation, ploughing, reseeding, tree planting and the application of artificial fertilisers or herbicides.

An area of grassland may have developed a special interest because of historical management. Management influences the plant and invertebrate communities, with some species restricted to grazed pasture and others to meadows cut for forage. Timing of mowing determines whether early-flowering or late-flowering plants thrive, while invertebrate communities are very sensitive to changes in grazing management. Consider the implications to existing wildlife interests before changing management. The isolation of many important grasslands today means that, if rare plants and invertebrates associated with a particular habitat are lost, they are unable to recolonise, even if the habitat is restored. When unimproved grassland was widespread, it would have been easier for species to recolonise if management made the site temporarily unsuitable.

It is important to identify the valuable species and features of an area of grassland and manage for them. Avoid management for particular species if the situation is inappropriate for them. For example, lapwings are unlikely to nest in small fields next to woodland, even if the ideal sward is created and food abundance is high, because they generally select open fields to enable easier detection of predators. Clear and appropriate targets are also essential when trying to restore or create a new habitat. Local and national Biodiversity Action Plans can be helpful in targeting species and habitats of conservation concern.

Restoration of species-rich grassland

It can be particularly difficult trying to increase botanical diversity in improved and semi-improved grasslands. High residual levels of fertility, particularly soil phosphorous, impede the restoration of such grasslands, even in the absence of intensive management. Grazing has little effect in reducing fertility, but annual cutting and removal of herbage reduces levels slowly. Significant levels of white clover in the sward will slow down nutrient depletion. The time taken for soil phosphorous to drop to levels associated with botanically diverse grassland is dependent on initial fertility, but can take several decades. The species that colonise the area as nutrient levels drop will depend on the content of the seed bank. The time and expense involved in such projects will be best justified where some of the following criteria can be satisfied:

- Existing soils are of low fertility.
- Areas are recently known to have been semi-natural grassland and are likely to have retained characteristic species in the vegetation or soil seed bank.
- Sites are adjacent to existing high quality habitat.
- Shallow soils or steep slopes reduce the nutrient-holding capacity of the site.

Many of the principles in restoring diversity on improved grassland are shared with the creation of grassland habitats on arable land (see section 3.20).

Further reading

Crofts A and Jefferson R G (1999) *The Lowland Grassland Management Handbook*, 2nd edition. English Nature/The Wildlife Trusts. English Nature, Peterborough.

3.20 Arable reversion to grassland and heathland

- When considering arable reversion to grassland clear and appropriate objectives are essential.
- A soil test will indicate the potential for reversion to species-rich grassland.
- When creating wet grasslands, it is important to establish a good soil structure before changing water levels.

3.20.1 Wildlife use

In many parts of the UK, Grassland habitats of high conservation value have become rare and fragmented, leading to a decline in species dependent on them. The reversion of arable or temporary grassland into native grassland habitats in suitable locations will play a part in reversing these losses. Local and national Biodiversity Action Plans are helpful in targeting species and habitats of conservation concern and funding may be available from agri-environment schemes.

Biodiversity Action Plan habitats that could potentially be re-created through arable reversion include:

- coastal and floodplain grazing marsh
- lowland calcareous grassland
- lowland acid grassland
- lowland heathland
- lowland hay meadows
- purple moor-grass and rush pastures.

Birds and other wildlife benefit from mosaics of habitat such as those provided by mixed farming. Where arable farming dominates, the reversion of small areas of arable land to grassland can provide important nesting and feeding habitats for some species (Plate 3.29).

As well as wildlife conservation, reversion to permanent grassland can provide other benefits. It can be used to improve landscapes, protect underlying archaeology and facilitate public access. Water quality can be improved through reverting areas vulnerable to soil erosion. The creation of wet grasslands can help to reduce flooding elsewhere in the river catchment.

3.20.2 Practical management

Factors that will determine the potential for recreating a particular grassland habitat include the location, geology, soil type and drainage. It is important to have clear and appropriate objectives for a site and ensure that the management required after reversion is feasible. Arable reversion is a long-term commitment requiring patience to achieve the desired outcome.

Reverting cultivated land adjacent to existing habitat to extend and link up areas has several benefits. It offers a better chance of flora and fauna associated with the habitat colonising the site. Seed can disperse with livestock, machinery and flooding from adjacent areas. It may also aid management of the existing area. Larger grazing areas may facilitate more economically viable grazing or better control of water levels in wet grassland habitats.

Chris Gomersall/rspb-images.com

Plate 3.29
Flower-rich grassland created on old arable land.

As the land to be reverted may have been under arable cultivation for many years, the soil fertility and pH may not be conducive to some grassland communities. A soil test will indicate the soil's pH and fertility levels. Ideally, soils suitable for reversion to species-rich grassland will have a soil phosphorous index of zero or one. Low yielding land due to poor soils or drainage are often the most suitable for reversion, being more likely to be ecologically and economically viable.

Grassland can be established through natural regeneration if there are still viable seeds of the desired plant community in the seed bank. It is the cheapest method of establishing a sward, also guaranteeing the local provenance of the seed. However, results can be unpredictable. The type and viability of seed in the soil is difficult to ascertain. It will depend on the length of time the field has been in arable cultivation, the intensity of management during this period and whether the preceding grassland contained the desired flora. Ideally, the site should have supported native grassland within the last decade. Allowing the field to germinate in the spring offers the opportunity to find out the quality of the seed bank. If the establishing vegetation is unsuitable, the field can be re-worked and seed sown in the autumn. Natural regeneration on set-aside may result in good quality grassland that can be turned into arable reversion.

Sowing a seed mixture minimises weed infestations and ensures the quick establishment of a sward. It also allows the sowing of desired species. A wide variety of seed mixtures can be purchased but the seed of some plants are unavailable or particularly difficult to establish. Seeds vary from agricultural cultivars, usually grown overseas, to mixtures harvested from existing semi-natural habitats in the UK. The latter is desirable, but can be significantly more expensive.

The spreading of green, unwilted hay from a suitable donor site onto the area proposed for reversion is an alternative method of reseeding. It can be relatively more cost effective than trying to source local provenance seed, though germination rates can be unpredictable. The donor site should be carefully chosen, matching the characteristics of the reversion site as closely as possible. The donor field should be cut when the main grasses still hold ripe seed (traditionally-managed species-rich hay meadows should be cut at the normal hay-making time). There are various ways of moving the green hay, such as picking it up with a forage harvester and spreading with a muck spreader. Green hay can also be baled, moved and spread. Fields need to be close together to prevent green hay heating up and seed deteriorating. Round bales should be moved and spread within a couple of hours of bailing.

Adequate soil moisture after sowing is vital for successful establishment, making late summer and early autumn the best period to sow. Cultivations before sowing should achieve a fine and firm seedbed. If pernicious weeds are likely to be a problem, they are best tackled by herbicides in the preceding crop. Short periods of intensive grazing or regular cutting during the first growing season will help to suppress annual weeds and prevent a few species dominating. Grazing is preferred as it improves soil–seed contact, increases tillering and reduces competition of vigorous plants earlier. Grazing also allows light into the base of the sward, which helps seed germination. Perennial weeds such as creeping thistle should be spot treated or weed wiped. In subsequent years, management will depend on the objectives for the site. During establishment, a new grass sward can be low yielding, but over time, productivity should improve. Former arable soils can be particularly low in organic matter and the application of manure can help sward establishment and improve soil structure.

Lowland wet grassland

Wet grasslands can provide suitable breeding habitat for a number of wading birds such as lapwings and redshanks. Waders require soils rich in invertebrates and moist enough through the spring for invertebrates to be easily accessible to the birds and their chicks.

Trying to revert an arable field to grassland at the same time as changing the water levels can prove problematic. It can result in weak soils of poor structure, low in organic matter. Such soils have fewer earthworms and dry out earlier in the spring, reducing access to those invertebrates present. Management is also more difficult as the period when soils can support grazing or machinery is shortened. It can also result in poor sward establishment.

When creating wet grassland from arable land it is important to establish a good soil structure before changing the hydrological regime and re-wetting the area. If the soils coming out of arable use are compacted then this problem needs to be resolved. Subsoiling may help, but will not generate good structure alone. Establishing a deep-rooted sward, raising the level of organic matter and increasing the activity of organisms like earthworms and fungi are critical to achieving a good soil structure. These aims will be best achieved with a period of dry grassland management. The addition of organic matter through manure in this period will help increase earthworm populations and improve soil structure. A dry period of two years may typically be needed to create soil conditions capable of tolerating wet grassland management but will vary with different sites.

The botanical diversity of wet grasslands varies considerably. Initial soil fertility and pH are therefore less of a problem than they are in the creation of more sensitive grassland communities. Once re-wetting begins, the sward will change to species more tolerant of wet conditions. These species could be included in the original mix, drilled into the existing sward later or allowed to colonise naturally. Grasses tolerant of wet soils include rough-stalked meadow-grass, Yorkshire fog, creeping bent and tufted hair-grass (which should be used sparingly). Ongoing management should aim to maintain a good soil structure by avoiding prolonged saturated conditions and ensuring that soils are not too wet when using machinery or during periods of grazing (see section 3.21).

Heathland

Lowland heathland has become highly fragmented during the last 100 years, in part due to conversion to farmland. Reversion is both feasible and highly desirable on land abutting existing heathland. Techniques for reversion need not be very expensive, but may be time consuming. The objective should not be for total heather cover; a mosaic of heather, gorse, acidic grassland and scrub will have greater wildlife benefits and take less time and expense to create. Woodlarks, skylarks, stonechats, linnets and yellowhammers are all birds of conservation concern that will benefit. Soil nutrients will need to be reduced to very low levels and pH to less than 5.5 for reversion to be successful. The vegetation will change as fertility declines, and will develop into an interesting acidic grassland sward with low-input grazing. Heather can be introduced by seed harvesting from the neighbouring heathland.

Grass pockets in arable landscapes

Grass pockets can be put back into arable landscapes through establishing small areas of grassland on the least profitable soils. These could include margins along field boundaries, awkward field corners and islands within the field that have difficult slopes or where yields are lower due to shallow or damp soils. Invertebrates benefit from the lack of cultivation, making grassland areas valuable foraging habitats, particularly for birds in the breeding season. For example, lapwings will nest in spring arable crops, but their chicks need a foraging habitat rich in invertebrates such as a short, damp grass sward (lapwings generally use grazed grassland). Grass margins provide nesting habitat for species like the grey partridge and yellowhammer. Skylarks and corn buntings may use in-field areas. However, soil containing rare arable weeds require regular disturbance and should be managed for these plants rather than reverted to grassland. In pastoral landscapes, pockets of arable land are vital in maintaining populations of several seed-eating birds. The reversion of arable land to grassland may be detrimental in such areas.

Further reading

Peel S (2001) Guidelines for the reversion of arable land. *Enact* 9, No 4.

3.21 Re-wetting and wet grassland management

Ensure that the site is suitable for the target species, is capable of maintaining the required water regime, can be managed appropriately and that there are no conflicting interests before starting a re-wetting project. Re-wetting should be undertaken gradually to assess the response and handle any problems that arise before managing on a larger scale.

The ideal water regime is to flood in winter, maintain high water levels in spring, but to dry out to allow management and recovery in summer. The sward of wet grassland is managed by grazing or mowing; cattle grazing creates the ideal sward for most bird species.

3.21.1 Wildlife use

Several species of wetland bird of conservation concern, notably waders, breed on farmland. They may nest in wet grassland, in-bye pasture or marshy areas. However, drainage and agricultural improvement of grassland to provide better grazing and forage has greatly reduced suitable areas for feeding and nesting birds. The ditches associated with these wet grasslands can support a rich variety of wildlife, including aquatic vegetation and invertebrates such as dragonflies, water beetles and snails, as well as declining species such as the water vole.

An opportunity to offset some of the declines in wetland wildlife on farmland is available by raising water levels, or 're-wetting', in order to re-create wetland habitats (Plate 3.30). Wading birds have specific requirements for nesting and feeding, and their chicks require a supply of high protein invertebrate food during the breeding season. Suitable habitat may also be created for yellow wagtails and reed buntings, while ditches may attract feeding water rails and grey herons.

A rich supply of insects will also help other birds, such as the tree sparrow, which rely heavily on insect food for their chicks.

Existing wet grassland is a very valuable habitat for a range of plants and animals including over 40 bird species of conservation concern (including wintering and breeding wildfowl and waders – see Plate 3.31). Around 500 species of plants have been recorded on wet grassland and associated ditch habitats.

R Winspear (The RSPB)

Plate 3.30
Land restored to wet grassland, providing habitat for wintering wildfowl and breeding waders such as the lapwing, redshank and snipe.

R Winspear (The RSPB)

Plate 3.31
Unimproved wet grassland in north Norfolk provides nesting habitat for lapwings and redshanks – the varied topography means that the hollows can hold water earlier in the spring.

Table 3.15 Key issues to be considered before re-wetting farmland.

Issue	Rationale	Points to consider
Geographic location	Breeding waders have been lost from large areas of the country. Newly created habitat may be only slowly colonised by target species	• What are the target species? They should ideally be present in the locality to enable colonisation
Site location and size	Surrounding habitat, site size and access are important in achieving successful restoration	• Is there a history of wetland habitats in the locality? Colonisation will be quicker if adjacent to other wetlands • Is the site large enough to meet the objectives? Larger sites will attract more species and offer greater flexibility • Can the appropriate grazing or cutting management be provided to maintain the ideal habitat? • Is the surrounding land use appropriate? Ideal sites will be unenclosed and relatively free of hedgerows and trees
Hydrology and soils	Control of water is critical to achieving results. Adult waders and their chicks feed in damp soil and shallow water with muddy margins	• Are the soils suitable? Free-draining soils are generally unsuitable unless the water table is close to the surface • Can water levels be maintained throughout the spring and early summer? (water control structures will be needed in many cases) • Is there a clear understanding of the water flows entering and leaving the area? • Are land levels and field topography suitable? • Does altering the drainage affect other land? Consult with appropriate authorities to ensure there is no conflict
Potential conflict with other features	Re-wetting should *not* be considered where there is a conflict of interest, for instance where there is an existing environmentally, historically, archaeologically or culturally important landscape	• Does the land have existing conservation value, eg a species-rich flower meadow, or use by other birds (corncrake, black grouse), that may be damaged by raised water levels? • Is the land a Scheduled Ancient Monument, other archaeological site, or ridge and furrow field system? • Are there existing public rights of way?

3.21.2 Practical management

The wetness of an area of land through the year is the result of the interactions between rainfall, summer moisture loss through evapo-transpiration (combined loss from both vegetation and open water surface) and the soil/geological conditions. In natural situations, the resulting vegetation, be it wet grassland, marsh or swamp for example, will be the community of plants best suited to the conditions. Where re-wetting is considered, the aim is to combine water control and appropriate land management to produce the desired wetland habitat, usually wet grassland. Re-wetting should only be attempted in suitable areas and it is therefore important to consider all the issues before proceeding. Where necessary, specialist advice should be sought. Table 3.15 identifies the issues that need to be considered.

Site history

Many farms will have held patches of wet grassland in the past, but most will have since been drained. Where habitat enhancement is proposed, the area targeted should generally have been historically wetter. Knowledge of the past management history, notably drainage, will be valuable and in some cases, simply reversing or controlling the outflow of water from a drainage ditch will bring about the desired result.

Knowledge of habitats and species present in the locality, both past and present, should be used to inform decisions on the desired end result. Where the farm still has grasslands that hold high water levels throughout the spring, then simply applying the appropriate management should ensure that they provide the right habitat structure for breeding waders. It would not be desirable to change the land use on existing wet pastures, but re-wetting may enable favourable conditions to be extended. If starting from arable land see subsection on lowland wet grassland in section 3.20.

Understanding soils

Understanding how water will behave in a particular soil type and the rate at which it will move through the soil (the hydraulic conductivity) is important in any re-wetting proposal. Water moves faster through soils composed of large particles (sand or peat) and well-structured soils (eg some clays which have not been ploughed). Water moves slower through soils composed of small particles (clay or silt), particularly where the soil structure is poor. Some clays may be virtually impermeable. Weathering, plant root action and soil invertebrates will, over time, act on unstructured clays to create small cracks, increasing water retention and hydraulic conductivity in the process.

When considering drainage, and therefore re-wetting, soils may be grouped as follows:

- *Free-draining soils* are permeable, they are unlikely to have been wetland in the past, or to have been drained. They are mostly composed of sands or well-structured clays. Wetland habitat creation is unlikely to be desirable or successful on such soils, at least not without a great deal of effort.
- *Groundwater gleys* (soils permanently or periodically waterlogged) are usually composed of similar particles (sands, peats and well-structured clays) and are also permeable, but may remain wet all year if there is a high groundwater table. In general they may be made wetter by installing water control structures, such as sluices, to retain high groundwater levels.
- *Surface water gleys* tend to be soils with a low permeability, such as those composed of clays and silts. They are often waterlogged in the winter but dry out in summer. Gleys in an arable situation would usually be drained with tile drains emptying into a ditch and it is possible to increase wetness by blocking outflows. This will raise water levels in winter and retain higher levels into spring. However, additional water will usually be required to retain summer wetness.

Assessing the soils within the project area is desirable. This may be achieved by digging a pit of one to two spade widths and one to two spade depths and looking for the different layers of soil present, the structure and composition of the soil in each, and the presence of the water table. This should be repeated at several locations to assess the variability of the soils.

Topography

A basic requirement for a re-wetting project is a field with the right topography to hold water, and therefore an understanding of the topography of the proposed land for re-wetting is a vital component of success. On a large scale, a topographic survey measures the height of ground level, water level and water control structures, thus providing information on slopes, flows and likely areas of flood. This may be achieved using sophisticated levelling equipment. However, the simpler method of recording a 'winter puddle' map, perhaps assisted by the placing of a few sandbags in a ditch to allow water to back-up, will allow the water to do the levelling work.

On a smaller scale, ground-level variation within a field can affect the plant community and water regime in directly adjacent areas, for example within ridge and furrow grassland, or on land with a complex pattern of rills. Such variation, best judged by eye, is extremely important in achieving ideal conditions for wet grassland species. Fields with a varied topography will enable the creation of a mosaic of wet feeding areas, drier nesting areas and a gradation in between that will dry out at different times throughout the spring. This is also important, as some crucial invertebrate food sources, such as earthworms and leatherjackets, will be lost in areas where the soil is inundated for long periods.

Water supply

Restoration of wetland habitats usually depends on raising water levels, so the availability of water is a crucial factor. There are a variety of potential water sources and the supply of water will be dependent on local climatic factors. Water input will depend on rainfall, surface water (river/stream flows) and groundwater levels. Water loss will depend on evapo-transpiration and seepage (groundwater seepage, drainage or runoff).

Groundwater sources may be utilised through the inflow of water from springs or seepages or by installing water control structures on outflows to retain the groundwater height. Surface water can be re-directed from ditches or streams to the desired location. Again, water control structures may be required. Consider any likely impacts created upstream by blocking or diverting drainage and consult with the necessary statutory agency (the Environment Agency in England and Wales, the Scottish Environmental Protection Agency in Scotland or the Rivers Agency in Northern Ireland) for further advice. In potentially complex situations, it is advisable to calculate a water budget using local climate and drainage data. In most cases, specialist advice should be sought.

Implementing the project

Any re-wetting should be undertaken gradually so that the response, both in terms of the habitat creation and the farming consequences, can be assessed and any problems addressed before it is undertaken on a large scale. Although larger areas are preferable (greater than 4 ha), there is no minimum size of project – even small-scale re-wetting by blocking a few field drains will provide some benefit. Snipe may nest in 0.2 ha of tussocky grassland with shallow pools, although predation rates in small patches are likely to be higher.

Once the key issues such as soils, topography, water supply and site history have been considered and proved satisfactory, the project can be implemented. The following sections look at managing and maintaining water levels, and creating features within the fields.

Managing water levels

Re-wetting may be undertaken in a range of soil and hydrological conditions and the simplest schemes are often no more complicated than reversing or reducing the drainage function in a particular area. In many areas with a wetter climate, the rainfall is sufficient to keep sites wet into June, and reducing the rate of runoff is sufficient. Assess the soils and drainage patterns for the area and if necessary, block any drains that take water away or redirect others to drain into it. Use of the least productive pastures will mean less of a loss for the farmer and may provide the best results.

In drier locations, it will be necessary to direct water to a chosen location and/or install water control structures to reduce losses. Alternatively, there may be an opportunity to create a bund (low bank) to retain water. Water control structures, such as sluices, enable the ability to control the inflow or outflow of water and thus the water level in the field. In the permeable groundwater gleys, sluices may be used to maintain groundwater levels and thus soft feeding conditions over the field. However, seepage is likely to be a key issue. In the less permeable surface water gleys, high ditch levels to maintain shallow surface flooding in low points in the field are critical. The provision of shallow water and muddy margins are important to feeding waders and without control the grassland may dry out too soon in early dry weather, while a wet spring may result in levels remaining too high.

The most cost-effective sluice is likely to be constructed with a length of plastic piping, either rigid pipe with a swivel end or flexipipe, laid through an earth dam in the outflow ditch or bund (Figure 3.11). Each end extends beyond the dam, and the upstream end is held at the desired level. Flexipipe will normally need weighting to keep the lip submerged and require a

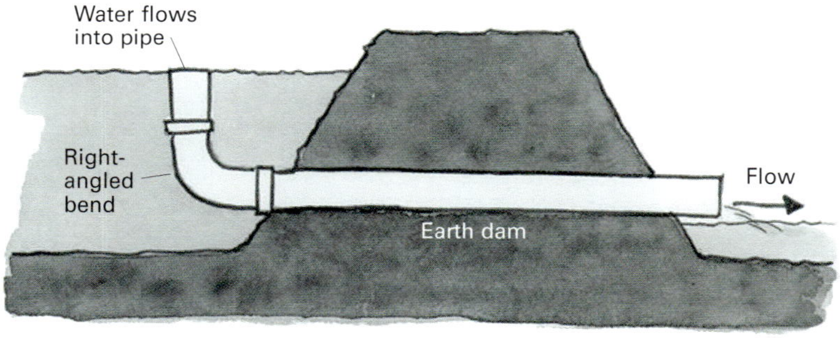

Figure 3.11
Diagram of a pipe sluice. Bunds need to be carefully engineered so that they are stable and impervious. Construction is easier in areas with loamy or clay soils and bunds should be keyed into an impermeable substrate. They should extend well into the bank on each side to reduce seepage losses and gentle profiles are advisable to enable mowing of any colonising rank vegetation. If a bund is necessary, aim to locate it appropriately within the landscape.

length of rope to hold the upstream end at the desired level. Adjusting the upstream end (by swivelling the pipe or raising or lowering the rope) will set the desired water levels. Other options are available, for example: drop-board sluices. In general, these are more costly in time and resources to install.

The water regime

The water regime is the combination of water table depth, the length of time that depth is maintained and at what time of year. Shallow pools within the fields are beneficial as they are ideal feeding habitats for all species. However, the presence of pools is only essential for redshanks. Maintaining high summer groundwater levels where possible will suffice for other species. Extensive and prolonged flooding may attract birds in the short-term but ultimately critically reduces the abundance and variety of soil invertebrates. In practice, in a field with variable topography and a high water table, surface water that covers a small percentage of a field (around 5%) will provide adequate open water and suitable soil conditions for a substantial proportion of the field (Figure 3.12).

The water level should naturally reduce slowly during the spring through evapotranspiration. Alternatively, depending on weather, let water out slowly to achieve the desired condition. Beyond July, the water table should be taken down to enable agricultural operations to be undertaken on the fields. A summary of a regime to meet the requirements of most species is as follows:

- *December–March:* high water table with 'splashy' conditions over the field and/or shallow flooding from 1 to 30 cm over up to 30% of the field. Allow water levels to fluctuate to avoid stagnation.
- *March–May:* high water table, at field level over 30% of the field and/or shallow flooding on 5–10% of the field.
- *May–July:* water table within 20 cm of field level on average, shallow pools shrinking with muddy edges.
- *July–December:* water table dropping to at least 40 cm below field level. No pools. Grassland dry enough for management.

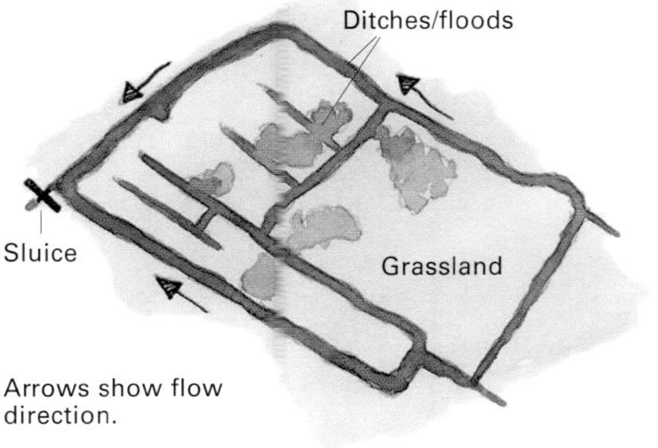

Figure 3.12
Diagrammatic representation of drainage channels and extent of early spring flooding in a wet grassland field.

Habitat management

Wet grasslands are the products of agricultural management, usually by grazing or mowing. Such management halts the natural succession to coarser vegetation types such as scrub. The timing of management, its intensity and frequency all influence the habitat. All resulting habitats will support wildlife, but they will be ideal for some species and wrong for others. The trick is to achieve the desired habitat for the target species.

Grazing with livestock, usually cattle, at a moderate intensity is ideal for nesting waders as it (a) creates a mosaic of tussocks and short turf used for nesting by a range of wader species and (b) augments the invertebrate population through dunging. To reduce trampling of nests, avoid grazing through the nesting period where possible. However, some grazing may be essential to maintain the desired sward structure, particularly where lapwings and redshanks breed. Heavy grazing from late summer onwards will restore the required sward heights for the following year. If grazing is not possible, cutting will be essential. Cutting should be timed for suitably dry periods after the end of the breeding season, usually between August and October. Expect to cut at least twice per year if grazing is not possible. Where re-wetting takes place on naturally damp land, rushes may increase rapidly and cause a problem if not managed (see section 3.23).

Liver fluke is a very debilitating internal parasite caused by the parasite *Fasciola hepatica*. It can affect all grazing animals but is primarily a disease of sheep and cattle. As well as grazing animals, liver fluke requires another intermediate host, the dwarf pond snail (*Lymnaea trucatula*). As this species favours wetter soils, liver fluke is more prevalent on wet grasslands. Livestock can be treated effectively with a flukicide, but control and prevention of liver fluke should form part of an animal health plan drawn up with a local veterinary surgeon. Avoiding grazing wet areas between October and December can reduce infection, as infestation is greatest in these months. It may also be more appropriate to graze wet grassland with cattle, rather than sheep, which are more susceptible to acute liver fluke.

Further reading

Benstead P, Drake M, Jose P, Mountford O, Newbold C and Treweek J (1997) *The Wet Grassland Guide: managing floodplains and coastal wet grasslands for wildlife*. The RSPB, Sandy.

Gilbert O L and Anderson P (1998) *Habitat Creation and Repair*. Oxford University Press, Oxford.

3.22 Creating wader scrapes and wet flashes

Scrapes or wet flashes can be created by reducing drainage from natural depressions, or by excavating shallow pools. Important features are shallow slopes, plenty of edge and deeper points sufficient to hold water through until June.

It is preferable to be able to control the water level so that it can be lowered gradually through the spring to expose fresh mud for foraging birds, and dried out in the summer to allow management such as rush control.

3.22.1 Wildlife use

An opportunity to offset some of the declines in breeding waders on farmland is possible through the creation of scrapes and wet flashes with sparse, marginal vegetation in grasslands with the potential to hold these species. These can provide important feeding areas for adult and young birds alike, particularly later in the spring.

The conditions created by the periodic flooding and drying of ephemeral water bodies attract a limited but specialised range of invertebrates. These often occur in very high numbers because of reduced competition and few predators. The water body is often rich in nutrients because of the levels of organic matter. This encourages high rates of invertebrate reproduction, particularly of midge and beetle larvae, which are a valuable food source for waders and their chicks. A rich supply of insects will also help other farmland birds such as reed buntings, tree sparrows, song thrushes and yellow wagtails.

As the water levels in the scrape are lowered, or dry naturally, the bare edges make the invertebrate food source more accessible. Moreover, annual plants will germinate on the margins, providing additional food and cover for chicks. The seeds they produce will accumulate around the edges of the pool and will provide winter food for wildfowl as well as a variety of finches and buntings that come to the shallow margins to drink and bathe. Where there are no conflicts with other priorities, scrapes can also provide important wet features on arable land, low lying coastal land, degraded old meadows and pasture, and uplands.

3.22.2 Practical management

Creating the scrape

Scrape creation should only be attempted in suitable areas. These are often in low-lying, poorly drained areas of fields, where as a result, crop or grass productivity is low. It is important to consider all the issues before proceeding, and where necessary, specialist advice should be sought. The key issues that need to be assessed are similar to those when re-wetting grassland (see Table 3.15 in section 3.21 on re-wetting grassland).

Scrapes may be located in a range of soil and hydrological conditions but most frequently will be on relatively level open land, preferably seeking a known damp area where water lies naturally on impermeable soils (Plate 3.32). Creating a

Plate 3.32
A small, newly created scrape.

G Woodburn (The RSPB)

scrape is often simply a case of reversing or reducing the function of drainage in a particular area, in others water may be directed to a chosen location. Assess the soils and drainage patterns for the site and if necessary, block any drains that take water away from the scrape area or redirect others to drain into it. Consider any likely impacts created upstream by blocking or diverting drainage and consult with the necessary statutory agency (The Environment Agency in England and Wales, the Scottish Environmental Protection Agency in Scotland or The Rivers Agency in Northern Ireland) for further advice. In potentially difficult situations, it may be necessary to assess rainfall against evapo-transpiration and volumes of water flow throughout the critical spring period, using local climate data from the Meteorological Office. Expert assistance may be required at this stage.

Several small flashes could be created instead of one larger one, and will provide more marginal habitat, but may also require more maintenance. Sculpting a convoluted, or sinuous, edge to the scrape will increase available feeding area and is likely to provide shelter in windy weather. Water depths in the scrape in early spring should typically be between 0–25 cm over half of the area and the remainder 25–50 cm.

Type A: dry area with impermeable soil. Water is retained by bunds and a control structure.

Type B: permeable soil with high groundwater. Surface flooding occurs in a depression as a result of a high water table.

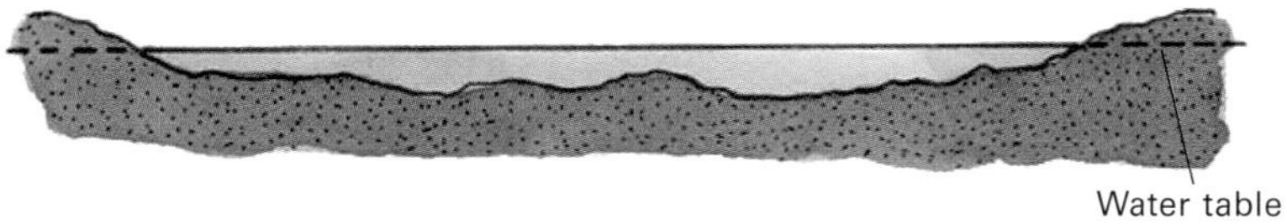

Type C: wet area with impermeable soil. Water retained in low-lying wet areas.

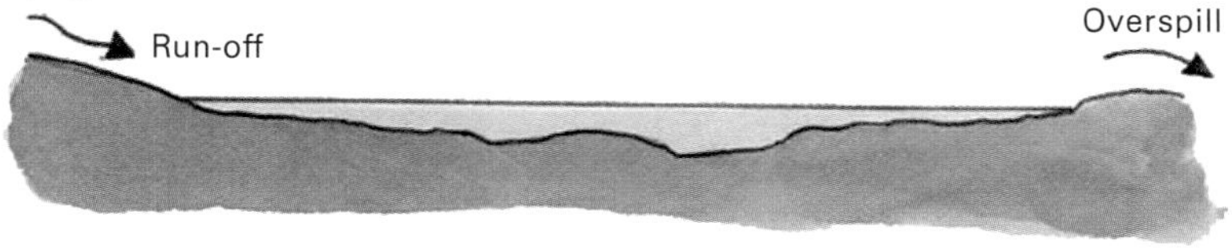

Figure 3.13
Types of scrape.

Ideally locate the scrape in a natural depression; otherwise, earthmoving, undertaken during a dry period, may be required to achieve the correct depth. A very gentle slope with an uneven finish will allow shallow, wet pools to remain longer within the scrape and allow a gradual exposure of the feeding surface. Different types of scrape are shown in Figure 3.13.

Any spoil material that is the by-product of excavating the scrape should ideally be removed away from the area. Alternatively, the spoil could be used to construct a bund around the downstream edge of the scrape. Note that this may limit the openness of the scrape and reduce its attractiveness to birds. Bunds need to be carefully engineered so that they are stable and impervious. It is important to consult with the appropriate authorities to check whether there are any restrictions on size or capacity.

Managing water levels

The provision of shallow water and muddy margins are important to feeding waders, and ideally, the water levels in a scrape should be controllable. Without the ability to control the inflow or outflow of water, the scrape may dry out too soon in early dry weather, while a wet spring may result in levels remaining too high. A simple water control device, or sluice, can be installed to help manage levels. The most cost-effective sluice is likely to be constructed with a length of plastic piping, either rigid pipe with a swivel end or flexipipe (see section 3.21).

The scrape should naturally reduce in depth slowly during the spring through evaporation. Alternatively, depending on weather, let water out of the scrape slowly (1 cm depth at a time) to create a muddy fringe. If the sluice is not connected to an existing watercourse, a soakaway will need to be created behind the dam to take the water drawn off through the sluice.

In Figure 3.14 the outer line represents the extent of the open water in early April, ideally surrounded by short grassland with up to 20% tussocky grassland. The

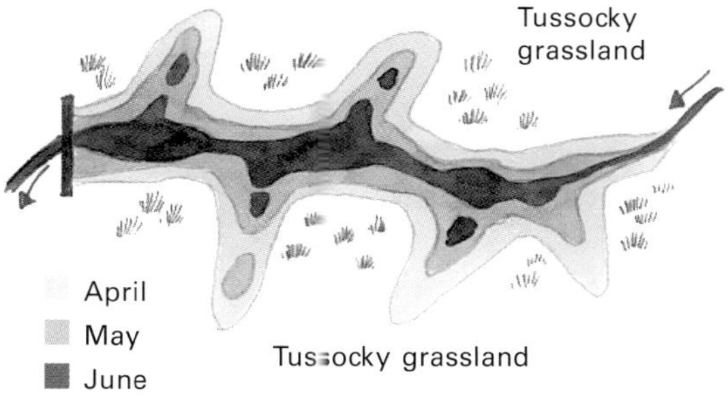

Figure 3.14
Hypothetical scrape showing receding area of water throughout spring and early summer.

middle line represents the shrinking area of water by the end of May and the inner by the end of June. Annual weeds will have grown on the mud and set seed. By August the scrape should be all but dry and ready for management.

Maintenance

It will be necessary to manage colonising plants, such as rushes or grasses, to prevent them from choking the whole area. Patchy cover of marginal plants will provide cover for young chicks, but if this exceeds more than 25% of the scrape, then management should be considered. Grazing with livestock at a moderate intensity is ideal as it (a) creates a mosaic of tussocks and short turf used for nesting by a range of wader species and (b) augments the invertebrate population of the margin through dunging. If grazing is not possible, cutting could be used. Cutting should be timed for suitably dry periods after the end of the breeding season, usually between August and October. It is not necessary to remove the cuttings, as they will initially provide a source of seed food and later, as they decompose, a source of insect food for birds.

Following summer/autumn management, re-flooding in winter will kill colonising perennial vegetation such as grass. Annual weeds, which germinate each year on the muddy margins as the water retreats, are important as they provide a large supply of seeds for dabbling ducks as well as number of passerines such as yellowhammers, reed buntings and linnets.

Further reading

Benstead P, Drake M, Jose P, Mountford O, Newbold C and Treweek J (1997) *The Wet Grassland Guide: managing floodplains and coastal wet grasslands for wildlife*. The RSPB, Sandy.

Gilbert O L and Anderson P (1998) *Habitat Creation and Repair*. Oxford University Press, Oxford.

3.23 Rush management

Rushes are a natural component of many wet grasslands, but certain species can infest if fields are neglected or the sward is broken through poaching or reseeding. If rush cover exceeds 30% of a field's area, its value to breeding waders is reduced. There are a number of options available to reduce rush infestations (see Practical management, below). Good sward husbandry reduces opportunities for rush establishment.

3.23.1 Wildlife use

Damp grassland on farmland is a very important breeding habitat for lapwings, curlews, redshanks and snipe. Rushes are often a natural component of such grassland. However, sometimes, certain species of rush can infest a field to such an extent that waders will be put off breeding in otherwise excellent habitat. Rushes can become particularly prevalent if fields are neglected or the sward is broken through poaching or reseeding. Exposed soil allows rush seeds in the seed bank the chance to establish.

Waders require variety in the sward structure. Taller areas provide cover for concealing nests and chicks while shorter areas are favoured for feeding. In fields with little plant diversity, rushes may be the only taller vegetation present, making them an important feature of the habitat. However, rushes are not essential for breeding waders. The required variations in sward structure may be provided by general plant diversity and a suitable grazing regime.

As a general guide, once rush cover exceeds 30% of a field's area, its value for breeding waders is reduced. Lapwings, which need good all-round vision, will be put off at lower levels of infestation than curlew and snipe. As well as reducing a field's value to waders, rush infestation will significantly reduce the grazing value of a field (Plate 3.33).

3.23.2 Practical management

There are over twenty species of rush in the UK, some of which can be localised and of particular ecological importance.

R Winspear (The RSPB)

Plate 3.33 A rush-infested field in need of management.

Four species that can be invasive and troublesome on farmland are the tussock-forming soft and hard rushes, and the creeping rushes: articulated and sharp-flowered rush. Perennials of damp and waterlogged fields, these rushes spread though rhizomes and prolific seeding (13,000 seeds/flower head), making them rapid colonisers of disturbed habitat. Rushes are tolerant of a wide pH range (eg soft rush 3–7, articulated rush 4.5–9) and moderately tolerant of annual cutting, grazing and trampling. Creeping rushes die back in winter and will be eaten more readily by stock than the tussock rushes, making them easier to control by grazing. They are also less tolerant of cutting, being readily controlled by a single late-summer cut.

Deciding on whether management should be carried out and the most appropriate method will be influenced by:

- site objectives
- botanical diversity of the sward
- feasibility of different options due to the sites wetness and topography
- relative costs.

Management will be an ongoing process, based on good sward husbandry and the methods described below.

Reseeding

There may be a temptation to cultivate the field and start with a new sward, particularly where there is little botanical interest. However, the ploughing and reseeding of damp grassland in the past has been the catalyst for rush infestation in many fields. Establishment of the sown sward is slow in an unfavourable environment, offering the opportunity for the millions of rush seeds in the seed bank to establish. Attempting another reseed is likely to incur the same problems. Cultivations associated with reseeding may also be damaging to archaeological features.

Cutting

Cutting rushes after the last wader chicks have fledged is an effective first step in managing rushes (Plate 3.34). The earliest timing will depend on the birds present. Snipe are the latest to breed, but should have fledged by August. Removing cuttings from the field is desirable, particularly where a lot of vegetation has been cut. Cuttings can mulch down to create new niches for rush regeneration.

Cut rushes as low as possible for the best results. Drum mowers achieve a very low

C Tomson (The RSPB)

Plate 3.34 Mowing in August to control rush invasion.

cut, but on rough terrain more robust machinery such as flail mowers will reduce wear. When flailing rushes it is important not to cut too low and scalp the sward, as this will stimulate germination of the seed bank. A single cut will have little impact on rushes and should be followed by one or a combination of grazing and herbicides (see below). Cutting rushes again around four to eight weeks after the first cut will help reduce rush cover in the following year. Leave some vegetation uncut around scrapes and very wet areas to maintain a mosaic of different habitats.

Grazing

In some circumstances, a single cut followed by grazing may be sufficient. Cattle are generally better than sheep at suppressing rushes. Creeping rushes are more readily grazed than tussock rushes. Stock should not be held on fields after they have grazed off non-rush vegetation. As rushes have low nutritional value, stock will loose condition. It also increases the likelihood of poaching.

Herbicides

Herbicides are perhaps best used when restorative measures are required in heavily rush-infested fields. MCPA and glyphosate are two approved chemicals very effective in managing rushes. However, both are broad-spectrum herbicides that will also kill other non-target plants. MCPA kills many broad-leaved plants while glyphosate is a non-selective herbicide, destroying all vegetation in the sward.

Boom spraying MCPA can be effective in managing rushes. However, this method should only be carried out where there is no botanical interest in the sward. It is also damaging to aquatic habitats. A more appropriate method of applying herbicide is through a contact applicator, such as a weed wiper. There are a variety of designs now available. Advantages of weed wipers include the following:

- Targeted application of the chemical avoids aquatic habitats and non-target vegetation.
- A low volume of herbicide and water is used.
- A weed wiper can be towed with a quad bike, reducing poaching in wet areas.
- There is no spray drift, allowing a large working window and safer application.

A significant height differential is required between the rushes and surrounding vegetation so that herbicide can be easily applied to rushes without killing non-target vegetation. This is best achieved by grazing the field quite heavily prior to weed wiping. Mature rushes can be weed wiped, but rush re-growth after topping is more susceptible to herbicide.

On large tussocks, the weed wiper may not make contact with the 'underside' of the tussock, requiring a second application. However, if insufficient time is allowed for the chemical to dry, herbicide may get onto machinery and spread onto the grassland. An experienced operator is essential on species-rich sites to avoid non-target vegetation.

3.24 Field margins and corners in grassland

As well as buffering existing wildlife-rich areas, field margins and corners can develop into important wildlife habitats in their own right. Leaving the margins of mown fields uncut creates rough grassland rich in invertebrates, seeds and small mammals, benefiting many birds. The establishment of legumes and herbs provides pollen and nectar.

3.24.1 Wildlife use

Field margins and corners can be used to buffer existing wildlife-rich habitats such as hedgerows, watercourses, walls, damp corners and areas of species-rich grassland from the impacts of management on adjacent farmland such as fertiliser and pesticide applications. They can also be used to create tussocky grassland that harbours large numbers of invertebrates, or to create flower-rich areas for the benefit of pollen and nectar-feeding insects, and provide a valuable foraging habitat for insectivorous birds and mammals.

Many farmland birds such as the song thrush, sparrows, finches and the yellowhammer favour feeding at the margins of fields. However, some species like the skylark actively avoid field margins. Invertebrate abundance and diversity is often greatest at the field edge due to the proximity of hedgerows, ditches and other habitats. Less mobile species remain close to the boundary all year. Sensitive management along the edges of fields will benefit these invertebrate communities and the birds that feed on them

Rough grassland along field boundaries provides a suitable nesting habitat for several birds. Rough grassland situated against a short, dense hedge provides an ideal nesting habitat for whitethroats and yellowhammers. Tall vegetation along damp areas such as ditches will provide nesting habitat for the reed bunting. Grasses and broad-leaved plants that are allowed to seed provide valuable food for seed eaters. For several birds, the availability of winter seed food is often the most limiting factor on modern livestock farms. Tussocks and dead vegetation will be colonised by small mammals that are in turn hunted by kestrels and barn owls.

Grass margins and field corners occupy the least productive area of the field where factors such as shade, compaction and less efficient input coverage generally reduce yield.

3.24.2 Practical management

Simply having a buffer to protect sensitive habitats from fertiliser and pesticide drift will be beneficial. However, the greatest benefits will occur if the margin and field corners can be managed to create habitats in their own right. Tussocky grassland and nectar-rich areas are two habitats that can be created. Margins and field corners can be established from the existing vegetation if the sward is relatively unimproved. If it is agriculturally improved, a better margin will be created by resowing with native grasses (see section 3.9), or desired plants integrated into the existing sward. Once established, retain the margin or field corner if the remainder of the field is cultivated and avoid fertilisers and pesticides drifting onto them (Plate 3.35).

Margins can be any width, but ideally, a variety of grass margin types between 2 m and 6 m will benefit the widest range of wildlife. Margins can also be of variable width to straighten up field edges and make field operations easier. Field corners may vary from simply rounding off awkward corners to encompassing a significant area of unimproved or damp pasture within a larger field. Margins along field boundaries are particularly useful in assisting the movement of less

D Buckingham (The RSPB)

Plate 3.35 A margin left uncut in an improved grass ley.

mobile species, such as plants or invertebrates, around the farm. Habitat creation in field corners can be used to create a large enough block of habitat to support a population of small mammals or a breeding pair of birds such as reed buntings. Agri-environment schemes may be able to fund the management of field margins and corners.

Nectar-rich margins

Establishing clover in field margins and corners can be used to replace the use of fertilisers in these areas, as well as providing nectar and seed. Red clover is an excellent nectar source for bumblebees. It may be feasible to establish a variety of wild plants associated with unimproved grassland in less fertile margins. Light grazing or annually cutting the margin will prevent legumes and herbs from being smothered. Nectar-rich margins can work well in fields cut for forage or rotationally grazed as plants have a greater opportunity to flower. This is different to the ungrazed 'pollen and nectar mixtures' covered in section 3.8.

Tussocky grass margins

Leaving the margins of hay, silage and zero grazed fields uncut can create rough grass margins and field corners. Some grazing is beneficial, providing access to food resources, but avoid grazing out the tussocky sward structure to conserve insect populations. Ideally, choose fields that receive the least grazing, or erect permanent or temporary fences to protect the margins from heavy grazing. Manage by cutting or grazing every few years. If a pernicious weed problem develops, then the problem area should be topped or treated with a herbicide by spot treatment or a weed wiper.

Early cover for corncrakes

In the British Isles, corncrakes are largely restricted to the Hebrides, Shetland and western Ireland, and very localised otherwise. In such areas, early cover is an important habitat. Corncrakes prefer to spend all their time in tall vegetation at least 20 cm high. This provides them with cover and a place to find invertebrate food, such as beetles, snails, slugs and worms.

When corncrakes arrive back from Africa in April and May, tall vegetation is scarce in many parts of the British Isles. This makes them vulnerable to predators such as cats and mink. Female corncrakes need areas of early cover to hide their first nests.

Beds of iris, nettles, cow parsley, reed and reed-grass all provide good early cover.

Brambles, bracken and dense stands of grass or rushes with a mat of dead stems and leaves from previous years' growth are not suitable. Corncrakes will use marshy ground at the margins of wetlands but avoid areas with permanent surface water.

Creating early cover – dumping manure, old silage bales or soil containing tubers of iris or nettle rhizomes can encourage suitable vegetation for creating early cover.

Early cover should be protected from grazing in late winter and spring. This can be done by fencing off corners of fields (a 'corncrake corner') or, even better, corridors running along field edges ('corncrake corridors'). An alternative to fencing off early cover areas is to remove stock from fields with suitable vegetation between late February and mid-March. Fields should then be left free of grazing until at least 1 August.

Early cover areas should be at least 0.15 ha in size. If it is not possible to establish individual areas of this size, then try to link different areas of early cover by strips of tall vegetation at least 2 m wide. This allows corncrakes to move around without coming out into the open. This early cover should be established in association with a minimum of 7 ha of hay/silage to be cut after 1 August using corncrake-friendly mowing. For smaller crofts, these areas may need to be split between neighbouring crofts. If more than 7 ha of hay/silage can be cut after 1 August then the area of early cover should be increased in proportion.

Early cover areas should be established next to fields of grass that will be managed in a corncrake-friendly way – that is, not grazed or cut until after 1 August and using the corncrake-friendly mowing method (see Figure 3.10 on page 142). When the grass in adjacent fields is tall enough, corncrakes will move between early cover areas and fields, but they will continue to use early cover areas throughout the summer.

Is your early cover too dense? Corncrakes will not use early cover if they cannot move easily through the vegetation. If there is a mat of dead or woody stems at ground level then the early cover is probably unsuitable and needs to be thinned out. The easiest way to do this is to open it up to grazing in the autumn and early winter. This will probably be necessary every two to three years.

Late cover for corncrakes

Tall vegetation may be scarce late in the summer if most fields in the area have been cut or heavily grazed. At this time, chicks from the second brood and moulting adults will be unable to fly. Areas established to provide early cover will continue to provide cover later in the season too. Uncut or ungrazed grass or late-harvested cereal crops can also provide refuges for corncrakes late in the breeding season and prior to migration to Africa.

Further reading

Haysom K A, McCracken D I and Foster G N (1999) Grass conservation headlands – adapting an arable technique for the grassland farmer. *Aspects of Applied Biology* 54, 171–178.

Appendix 1 Nesting times of bird species likely to be affected by farming operations

a. Species predominantly found in farm hedgerows and scrub (and number of broods per year, *species predominantly found in field margins). Dark shading indicates the main breeding season and lighter shading indicates extreme dates.

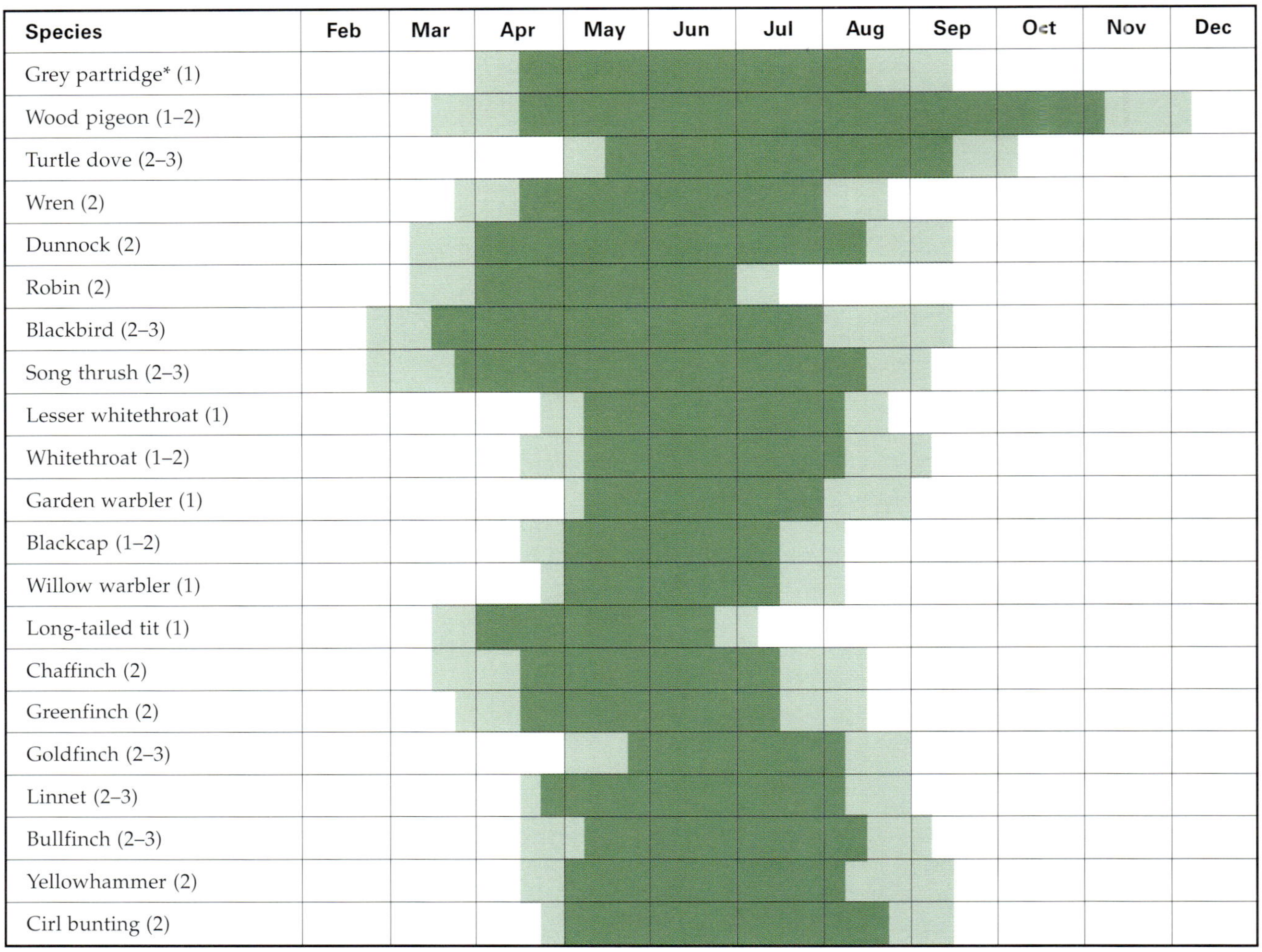

b. Species predominantly found in-field (and number of broods per year, a = species nesting in arable crops, g = species nesting in grasslands). Dark shading indicates the main breeding season and lighter shading indicates extreme dates.

Species	Feb	Mar	Apr	May	Jun	Jul	Aug	Sep	Oct	Nov	Dec
Marsh harrier (1) (a,g)											
Montagues harrier (1) (a,g)											
Corncrake (g)											
Stone-curlew (1–2) (a,g)											
Lapwing (1) (a,g)											
Snipe (1) (g)											
Curlew (1) (a,g)											
Redshank (1) (g)											
Quail (1) (a,g)											
Woodlark (2–3) (a,g)											
Skylark (2–3) (a,g)											
Meadow pipit (2) (g											
Yellow wagtail (1–2) (a,g)											
Whinchat (1–2) (g)											
Grasshopper warbler (2) (g)											
Reed bunting (1–2) (a,g)											
Corn bunting (1–2) (a)											

Appendix 2 Useful addresses

For information on agri-environment schemes:

Defra Rural Development Service
Regional Office details available from:
www.defra.gov.uk

Scottish Natural Heritage
12 Hope Terrace
Edinburgh
EH9 2AS
Scotland
Tel: 0131 447 4784
www.snh.org.uk

Scottish Executive Environment and Rural Affairs Department
Pentland House
47 Robb's Loan
Edinburgh
EH14 1TY
Tel: 0131 556 8400
www.scotland.gov.uk

Countryside Council for Wales
Plas Penrhos
Bangor
Gwynedd
LL57 2LQ
Tel: 01248 385500
www.ccw.org.uk

Department of Agriculture and Rural Development
Countryside Management Division
Annexe D
Dundonald House
Upper Newtownards Road
Belfast
BT4 3SB
Tel: 028 9052 0922
www.dardni.gov.uk

For farm conservation advice:

The RSPB
UK Headquarters
The Lodge
Sandy
Bedfordshire
SG19 2DL
Tel: 01767 680551
www.rspb.org.uk

Farming and Wildlife Advisory Group
National Agricultural Centre
Stoneleigh
Kenilworth
Warwickshire
CV8 2RX
Tel: 02476 696699
www.fwag.org.uk

The Game Conservancy Trust
Fordingbridge
Hampshire
SP6 1EF
Tel: 01425 652381
www.gct.org.uk

Scottish Agricultural College
Kings Buildings
West Mains Road
Edinburgh
EH9 3JG
Tel: 0131 535 4000
www.sac.ac.uk

Farming Connect
Y Lanfa, Trefechan
Aberystwyth
Ceredigion
SY23 1AS
Tel: 08456 000813
www.wales.gov.uk/farmingconnect/

Conservation Adviser
ADAS Boxworth
Battlegate Road
Boxworth
Cambridgeshire
CB3 8NN
Tel: 02890 520922
www.adas.co.uk

Index

Notes: *Italic* page numbers refer to photographs and **bold** numbers indicate cases, figures and tables.